WATER

PAUL

CARO
WATER

McGraw-Hill, Inc.
New York St. Louis San Francisco Auckland Bogotá
Caracas Lisbon London Madrid Mexico
Milan Montreal New Delhi Paris
San Juan São Paulo Singapore
Sydney Tokyo Toronto

English Language Edition

Translated by Patricia Thickstun
in collaboration with
The Language Service, Inc.
Poughkeepsie, New York

Typography by AB Typesetting
Poughkeepsie, New York

Library of Congress Cataloging-in-Publication Data
Caro, Paul.
   [*De l'Eau*. English]
   Water / Paul Caro.
      p. cm. — (The McGraw-Hill *HORIZONS OF SCIENCE* series)
   Translation of: *De l'Eau*.
   ISBN 0-07-009990-1
   1. Water. I. Title. II. Series.
GB661.2.C3613 1993   93-10537
553.7 — dc20

The original French language edition of this book
was published as *De l'Eau*, copyright © 1992,
Hachette, Paris, France.
Questions de science series
Series editor, Dominique Lecourt

TABLE OF CONTENTS

INTRODUCTION

In the last few pages of this short scholarly book, Paul Caro speculates on the daydreams inspired by water, a dominant theme in the human imagination through the ages. Water remains an enigmatic subject of research for both chemists and physicists, a volatile topic for the authorities charged with protecting the public health, and an object of anxiety for citizens threatened by drought and industrial pollution. The first thing that comes to mind about water is its status as an "element," as it has been considered since ancient Greece, and as we have continued to call it, in complete disregard of science, which, since Robert Boyle (1627–1691), has redefined the "element." Water was one of the first materials that the human imagination latched onto in the physical world to attempt to conform it to our desires. From this point of view, water has intrigued human cultures since well before the *physiologoi* of Ionia in the 6th century B.C. made water the subject of naturalistically oriented speculations. As a source of life and a means of purification, water is undoubtedly the prototype for spiritualized matter; the irrepressible attraction of water for human thinking was first expressed in religious or poetic form.

"All was water," according to Vedic tradition, as well as certain corresponding Taoist texts. Asia considered water to be the original "chaos," "the source of all things and of all existence,"—in Chinese *Wu-ji*. Within this as yet undifferentiated "raw material" reside all the possibilities of existence. The Polynesians, in identifying this Ocean of origins, this primordial substance, with the cosmic power itself, echoed the prayer of the Rig Veda:

> *You the Waters that comfort,*
> *Give us power*
> *Greatness, joy, vision!*

Babylonian cosmology also reserved a place for the chaos of water. Water occurs in two forms: the ocean of fresh water (*apsû*) on which the earth would later float, and the salty sea populated by monsters (*tiamat*). The poem of creation, *Enuma Elish*, begins:

> *When the skies above were not yet named,*
> *And below the earth had no name*
> *The primordial Apsû, which engendered them,*
> *Mummu, Tiamat, mother of all,*
> *Combined all their waters into one...*

The first words of Genesis, which continued to inspire heated debates among chemists until the 17th century, nevertheless imply that water came first: when God created the heavens and the earth, the earth was "without form and void; and darkness was upon the face of the deep. And the spirit of God moved upon the face of the waters." The existence of water thus preceded the first day. In fact, the entire Old Testament celebrates the magnificence of water.

To these cosmogonies correspond the fabulous qualities attributed to water. Life, vigor, and eternity reside in water. Living water or the "water of life," this celestial reality that so many cultures have reserved for the initiate, has magical medicinal properties. But all water, even the most ordinary, is revealed to be endowed with marvelous virtues: humanity has never ceased to dream of the curative virtues of springs, fountains, wells, streams, and rivers.

The timeless rituals of immersion proceed from the same idea: "Immersion in water," wrote the great Romanian historian Mircea Eliade (1907–1986), "symbolizes regression into the preformal, total regeneration, rebirth," adding that from the Islamic world to Japan, and including the rites of the ancient *fu-shui* Taoists (the "masters of sacred water"), ablution or aspersion plays an essential role in religious practice. He mentions that the Koran designates the holy water that falls from the sky as a sign of the divine, and that Islamic prayer must be preceded by ablutions. And Lao-tzu, the "Old Master" of Tao who lived in the 6th century B.C., teaches that water is the "emblem of supreme virtue." In the Greek world, statues of the great goddesses of fertility and agriculture were usually plunged into water. These sacred baths, which continued Cretan and Phoenician practices, were also known in several Germanic tribes.

Inheriting these rituals, Christianity sought to attribute a new significance to them. It achieved partial success in this endeavor at the price of fierce battles. The baptism of St. John brought not only healing of physical infirmities, but also remission of sins. Immersion in baptismal water, according to St. Paul (Romans 6:3), symbolizes the resurrection of Christ: through immersion, some of us shed our old lives, "die" and are "reborn," purified. This interpretation of baptism was the subject of a voluminous literature by the Church Fathers. In the Middle Ages, the French theologian Hugh of St. Victor (1096–1141) characterized water as the symbol of spiritual life: the soul is washed by the "waters of wisdom." Theologians refer to the Gospel according to Saint John (4:14) where Jesus said to the woman of Samaria: "whosoever drinketh of the water that I shall give him shall never thirst; but the water that I shall give him shall be in him a well of water springing up into everlasting life." However, this interpretation was not able to eliminate the popular devotion to water, a practice of pagan origin, despite the repression which began in the 4th century and continued through significant and repeated interdictions until the Council of Treves in 1227.

We can readily understand that under these conditions the knowledge concerning water met with powerful obstacles. Psychoanalysts have studied the feminine, maternal nature of the image of water in the unconscious. Gaston Bachelard, inspired by the major works of Carl Jung (1875–1961) on alchemy, found innumerable traces of this sexualization, not only in works of poetry, but also in scientific works of the past. He cites and comments on Stéphane Mallarmé (1842–1898):

O mirror!
Water chilled by boredom in your cold frame...

and on Paul Valéry's (1871–1945) Narcissus:

The slightest sigh
That I would exhale
Would rob me
Of what I would adore
In the blue and blond water
And the skies and the forests
And the rose of the waves.

Sexualized, water finds itself harboring ambivalences that are profound and enduring: water accompanies death as much as it does life, as was illustrated by Edgar Allan Poe (1809–1849) meditating alongside rivers and lakes. Heraclitus said: "It is death for the soul to become water." The poet echoes him when he presents water as an invitation to death. And Bachelard, evoking what he calls the "Ophelia complex," writes: "Water is the element of young and beautiful death, of death in bloom, and in the dramas of life and of literature, it is the element of death without pride or vengeance..."

One of the Ionian group of Greek philosophers who was the first to attempt to explain the world as the result of natural causes, Thales (625–547 B.C.) wondered about the origin of things. He categorized water as a primordial substance, conjecturing that the Earth floated on water, and thus explained earthquakes as the agitation of this subterranean ocean. Anaximander (610–547 B.C.) compared water to the Unlimited, and Anaximenes (6th century B.C.) compared water to air. However, according to the English historian Geoffrey E. R. Lloyd, Thales and Anaximenes posed different questions: Thales wondered only about beginnings, whereas Anaximenes questioned the intimate constitution of things by referring to processes still observable in natural phenomena. Beginning with Anaximenes, the notion of "element" (in Greek, *stoicheion*) took on the philosophical meaning that Aristotle (384–322 B.C.) transmitted to us: the principle of things. The ambiguity is nevertheless great: is it a matter of original substances or "simple" substances under which all compound things can be subsumed? We note the presence of this last concept in even the most ancient Greek thinking, as attested to by the Pandora myth told by Hesiod (mid-8th century B.C.) in his epic poem *Works and Days*: Hephaistos created the first woman by mixing and shaping earth with water.

But Empedocles (5th century B.C.) appeared to be the first in this line of thinkers to have superimposed the two versions of this notion: according to him, there are substances which are both original and simple. He designated earth, water, air, and fire by the name *rhizomata* ("roots"), and attempted to explain the entire diversity of things with these four roots whose combinations are governed by Love and Strife (*Philia* and *Neikos*).

Aristotle associated these elements with perceptible, tangible qualities, grouped by "opposites": dry and wet, cold and hot. The four possible combinations of these two pairs yield the four "elements": earth, water, air, and fire. Aristotle's influence on the human mind for the next two millennia is due to his having linked these ultimate constituents—whose existence was purely speculative—with sensory perception. The case of water is particularly enlightening. What happens, the philosopher asks, when water evaporates or when it is brought to boiling? It becomes "air," by the transformation of cold and wet to hot and wet. Conversely, when air condenses to become water again we are witnessing the substitution of cold for hot. Book IV of *Meteorologics* supplies many examples from which the power of conviction can be perfectly understood: not only did these examples correspond to observations that anyone could make, but they also took place in an explanatory system of the world which, from metaphysics to physics and from physics to natural history, presented an unfailing coherence.

Over the centuries, water thus led a double life as an "element" in human culture: as a key notion in religious practices referring back to the stories of creation and shaping human imagination by modulating original fantasies, and as one of the ultimate principles of explanation for learned consciousness. That between these two accepted notions there should have been a partial overlapping or complete identification is not surprising, as a world is involved where science was long kept under the frowning eye of religion.

Attesting to this, in the 16th century, is the powerful and controversial work of Bombast von Hohenheim (1493–1541), better known as Paracelsus, who thought of the Universe as an eternal river of life—a river spreads and breaks up into many streams, which meet, interact, and finally lose themselves in the ocean of their origin. In this cosmic view, the four visible elements are not, strictly speaking, "elements": they are only "bodies" giving perceptible form to real elements that are immaterial and dynamic and whose origin is unique. Paracelsus invokes the common experience of the people of his time: there exist, for example, beings who, unlike humans, are composed of only one element: gnomes of the earth, sprites of the water, and elves of the air. Magicians know how to call forth other "spirits" which are made up of two elements. Without yielding to the ridiculous temptation of making such spirits the precursors of modern chemists, we must nevertheless note that Paracelsus vigorously contributed to the "undoing" of the Aristotelian notion of elements. The Great Mystery (*Mysterium Magnum*), the Chaotic Egg from which the Universe was hatched and which directs its development, occurred by means of a primary fluid, fine and subtle, impalpable and invisible, that Paracelsus called Yliaster or Yliader, whose "condensations" and "coagulations" successively gave rise to astral matter, the heavens, and our own substance. This process was guided by the three primary forces that make up the world and its elements: Sulphur, Mercurius, and Sol. The four Aristotelian elements are the four modes of being of our substance, the four classes of material bodies. But in each real body, whether it belongs to only one of these elements or to a combination of several, there are the three "chemical" elements, forces formative of every bodily being.

The work of Paracelsus appears, as do all works of "alchemy," under the guise of a medical work, violently opposed to the medicine of his time. His influence extended until the middle of the 17th century, despite the persecutions endured by him as well as by his disciples, one of whom was Jan Baptista van Helmont, born in Brussels in 1577. After his encyclopedic studies at the University of Louvain, van Helmont's work incited controversies and was condemned in 1630 before he himself was arrested on March 4, 1634, interned, and kept under house arrest.

Like Paracelsus, van Helmont rejected the Aristotelian theory of the four elements, which he denounced as pagan. However, he remained fascinated by one of them: water. In contrast to Paracelsus, he proposed a return to the Bible, stating that the waters were obviously created before the first day and, even though they were not named, they were nevertheless included in the heavens; consequently, they participated in some way in celestial nature and, moreover, the superior waters were parents to the inferior waters, since they had been joined before their separation.

For van Helmont, water made up the material principle of all the bodies created by God. He thus proposed a grandiose water-based cosmogony. By means of learned transformation experiments, he believed that he had demonstrated and proven that "all bodies (which were believed to be mixed), of whatever nature, opaque or transparent, solid or liquid, similar or dissimilar (such as stone, sulfur, metal, honey, wax, oil, bone, brain, cartilage, wood, bark, leaves, etc.) are materially composed of simple water, and can be completely reduced to insipid water without the slightest trace of any earthly thing remaining." But how could the diversity of beings be explained if their substance was unique? By the action of a specifying ferment that penetrates water, a "formal and neutral being," for which he borrowed the name "aeon" from the 2nd-century Egyptian Gnostic philosopher Basilus Valentinus.

For van Helmont, water had its chemical apotheosis in the context of a world system which seemed to be resolutely anti-Aristotelian. Some disciples, ignoring this system, felt duty-bound to specify water's experimental elements. Thus the Danish scholar Alaus Borrichius attempted, without much real success, to use the stalactite model to explain the formation of stones from running water. But if this work was not engulfed in the "Museum of Horrors" (Gaston Bachelard) that represents the history of scientific doctrines, it is not only because van Helmont, as he went along, invented the word "gas" (from the German word *Geist*, spirit, or attributed by some to the Greek word *chaos*) to designate "aeriform substances." It was also due to the upheaval that van Helmont's work provoked, after that of Paracelsus, in reigning Aristotelian thought. This upheaval was the more effective because van Helmont was a contemporary of Galileo and Descartes.

In this sense, he paved the way for the "skepticism" of Robert Boyle (1627–1691), by the latter's own admission. It is undoubtedly to Boyle that we owe the first modern definition of a chemical element, as it appeared in his celebrated work *The Sceptical Chymist* (1662); it is technique that now defined an element as the limit of analysis. According to Boyle, water would thus not be an element. Nevertheless, this "truth" took more than a century to be understood and accepted.

This was the beginning of a very long and strange period in the history of chemistry. Experimental physics existed and was developing apace, but chemists who contributed to its development did not benefit from it. On the contrary, they made very little progress due to the reigning theories of physics—whether the Cartesian physics that inspired van Helmont or even Newtonian physics.

Perhaps this was due to the enduring fascination evoked by the old concept of elements as well as to the persistence of the many religious practices that referred to this concept. In the highly influential *Dictionnaire de Chymie* of Jean-François Macquer published in 1766, one century after Boyle, elements continued to be defined as "simple" bodies in themselves. "The bodies to which we have accorded this simplicity are fire, air, water, and the purest earth," wrote Macquer, although he added in a modern vein, without fully losing his composure: "Because, in fact, the most complete and most precise analyses that we are able to do so far have never produced anything else, after all, than one of these four substances." This explains the great success encountered in France by the *Hydro-theologie* [Theology of water, or essay on the divine goodness manifested by the creation of water] of Joannes Albertus Fabricius (1668–1736) after 1743.

This also explains the triumph of the first chemical system that gained recognition throughout Europe, the one proposed by the German chemist Georg Ernst Stahl (1660–1734), who was physician to King Frederick William I until his death. This system explained the formation of salts (neutral, acid, and alkali salts) as a combination of earth and water uniting by "affinity" with their fellow elements. Immanuel Kant, in his famous Preface to the second edition of the *Critique of Pure Reason* (1787), considered Stahl to be the founder of modern chemistry because he guided this discipline onto the "sure path of science," and ranked him above Thales, Galileo and Torricelli.

When the illustrious Professor Kant of Königsberg pronounced his judgment, four years had passed since the beginning of the Lavoisier revolution that was to destroy for good the Aristotelian doctrine of the elements. Neither Antoine-Laurent de Lavoisier (1743–1794) nor his contemporaries Joseph Priestly (1733–1804) and Henry Cavendish (1731–1810) were concerned with developing a system of Newtonian-inspired chemistry. Lavoisier responded to concrete demands of the authorities and referred to the empirical philosophy of Abbé Étienne Bonnot de Condillac (1715–1780). He presented to the French Academy of Science a paper on the composition of water in which he reported the results of a study he had carried out in the Vosges that discredited the theses of van Helmont on the transmutation of water into earth. In 1783, Lavoisier adduced experimental proof that water is a compound. The history of the chemistry of water could now begin.

It is this history of water that Paul Caro offers to the reader, who will certainly be surprised to learn the extent of the uncertainties and gaps in our knowledge of the water molecule (H_2O). Its disconcerting physical properties have led chemists to speak of the "bizarre behavior" and "abnormality" of this liquid that is so familiar to us. Old models are continually being replaced by new ones, and computer science has its full role to play. Nevertheless, quite a number of decisive questions about the "architecture" of water remain unanswered.

To these questions are added social and economic concerns that make water the business of the State, while issues of recycling and purity continue to preoccupy us. Clever copywriters know how to turn this to their advantage. Occasionally researchers get caught up in this fever. In the last few pages of this book, the author reminds us of a few ancient and recent episodes of an adventure in which water and dreams get mixed up even in the best of minds.

And so do sometimes dreams and profits....

Dominique LECOURT

I
LEGAL WATER

Hot or cold, water flows over the skin; cool or steaming hot, sweet or tart, it slides over the tongue and down the throat; it sings to the ear along the brook, it deafens near the waterfall; its great expanse fascinates our view; its reflection has a blinding, hypnotic power. Water has pleasant and unpleasant odors, scents and stenches, which it dilutes or concentrates; it surrounds the body in the bath, supports it, while at the same time the sly danger it represents demands alertness. Water in its manifold capacities touches the physiological: it awakens a wide range of sensations; contact with water and its presence stimulates receptors in all our organs. Water affects the entire gamut of our senses, which, for example, distinguish it from fire; its qualities and various forms can be perceived and measured by our natural tools of knowledge—the "primary receptors" that provide an immediate response, informing our brains about the physical state of the variables in our environment.

Water also speaks to the sensual—to pleasure, fear, and pain. Water colors our emotions, and is a frequent visitor in our dreams. Water is the most ordinary representation of liquid, or that which spreads, runs, wets, gushes, or floods. In anxiety as in sensual delight, it participates in an affectivity that links the physical and the mental when our emotions burst forth. Water influences our minds, our moods, our well-being. It insinuates itself into the many small events in our lives, into their most trivial aspects; into our daily concerns about rain and sunshine; into our subconscious pleasure of hands rubbing against each other under the spray of the faucet. We hardly ever give it a thought, unless some incident suddenly calls it to our attention: floods, interruptions in supply, droughts. Whether absent or present, water occupies a certain amount of our time every day. Most often just a simple commodity, it can invade our minds to the point of obsession, becoming a sort of goddess who demands hymns and prostration, the sacrifice of time. Water engenders folly, anxiety, or the innocent charm of a shared game.

The goal of this book is not to add a chapter to the theology of the joys and pleasures of water nor to establish a catalogue of everything that it can contribute to life, death, love, joy, or drama. I will attempt to stay within the realm of the perceptible without succumbing overly much to the charms of the sensual, in short, to speak of physical water as modern scholars understand it. To do this, it is necessary to take the path of realism and begin by accepting that, theoretically speaking, pure water does not exist! The definition of water is primarily bureaucratic—that is, it is the business of the government. According to Denis Diderot (1713–1784), who wrote an article on Hobbism in his *Encyclopédie*, Thomas Hobbes (1588–1679) suggested in the mid-1600s that "it should not be left to doctors or philosophers to interpret the laws of nature. It is the business of the Sovereign; it is not truth, but authority that makes the law." In fact, water is defined, meticulously described, and explored in all its complexity by the government, which pays scrupulous attention to it, not overlooking the slightest detail that could be the subject of complaints, disputes or doubts, not hesitating to change the rules when necessary, always adding new constraints. The government protects the throat and the body of the citizen, monitoring the quality of the product that is imbibed or is used to dissolve dirt, all the while imposing restrictions and "norms." Some figures will suffice to convince the reader of the necessity of the state's concern: for example, in the city of Paris, 179,600,000 gallons [680,000 m^3] of water are consumed each day, which is equal to 60 gallons [230 liters] per day per inhabitant. By contrast, in New York the figure is 180 gallons [680 liters] a day!

PURE WATER, DRINKING WATER, CONSUMABLE WATER

The French decree 89-3 of January 3, 1989 (amended by the administrative orders dated July 10, 1989; February 20, 1990; March 20, 1990; and March 11, 1991, as well as several bulletins), defines, in Appendix 1, the "quality limits" of the water provided for public consumption which must not be exceeded. "Pure water" is defined by a negative definition, by listing the maximum quantities of "foreign" substances that water can contain, thus setting the limits beyond which water must not be consumed. Natural mineral waters, spring waters, and bottled drinking waters are covered in decree 89-369 of June 6, 1989.

First there are the *organoleptic properties* of water, which theoretically can be assessed by natural instruments (*organon*, in Greek), namely, the senses: color, odor, and taste, to which is added evaluation of *turbidity*, which is defined as the "state of cloudiness of a liquid." Although we all consider ourselves capable of judging for ourselves, the law requires methods and laboratories to be used. Thus, color is measured by the platinum-cobalt method, which consists in comparing the sample with controls prepared from artificial solutions of a mixture of potassium chloroplatinate, crystallized cobalt chloride, and hydrochloric acid mixed in distilled water. The colors are compared by examining test tubes held upright against a white surface. Since it is necessary to have figures to express a judgment in science, the artificial solutions are classified in Hazen units, which correspond to a platinum content (one unit = 1 mg/L of platinum), and the color of the water being analyzed is measured by the figure matched by the control whose color it most closely resembles to the naked eye. The decree states that the color of water must not exceed the color of water containing 15 mg/L of platinum.

The question of odor is obviously much more difficult to resolve. An odor cannot be expressed by a number on a dial. There are no scientific instruments capable of measuring odors: the only known instrument is the nose. We must make do by defining a sensation perceived by the olfactory organ. Unlike color, which can be quantified by a spectrum of light absorption, odor does not have an independent existence. The nose is a powerful organ that uses a system of receptors to detect the presence of a certain concentration of molecules stimulating sensation, often a bouquet of active molecules. The brain analyzes this information in terms of pleasure or disgust, thus in terms of quality. It is therefore impossible to measure the odor of water without a sniffer. Since the law cannot do without figures, they are obtained by counting the number of operations required to make an odor disappear, as determined by a sensitive nose. The sample is diluted with odorless water which is obtained by "passing drinking water over activated charcoal pellets at a maximum rate of 6 liters per hour." If the water must be diluted a great deal, it is said that its odor is strong. An additional complication, of course, is that the sensation also depends on the temperature of the liquid; it is therefore recommended to work at two different temperatures: 54 and 77°F [12 and 25°C], says the law, so solutions need not be diluted beyond a factor of 2 at the lower temperature and a factor of 3 at the higher temperature. Specialists say that it is necessary to use water that has been heated to 104°F [40°C] or even 140°F [60°C] to make a reliable evaluation. By so doing, the odor of water can be classified in one of the following categories, which the reader will no doubt appreciate as examples of administrative poetry: "aromatic, balsamic, chemical, chlorinous, hydrocarbonous, medical or pharmaceutical, sulfurous, disagreeable (?), fishy, earthy, peaty, fecal, grassy, moldy, and muddy."

With regard to taste, the situation is hardly better: we are reduced to using our mouth and taste buds, while adding nasal and tactile sensations and taking into account temperature, pain, and other reactions. As in the case of odor, it is therefore necessary to dilute a sample with a "tasteless" water sample, although reference solutions whose taste has been calibrated, so to speak, are also used. These calibrated reference solutions are made with suspect substances (for example, chlorophenol, naphthalene, and chlorine), which allow a certain measure of quantitative evaluation to be made. The tasting ceremonial is carefully codified with regard to the way the taster "swishes" the liquid to be tested from side to side in the mouth, then balances a small amount of it on the taste buds of the tip of the tongue before slowly swallowing it. For the sake of accuracy, the task is given to panels of three "tasters" who have thoroughly rinsed their mouths with a reference water. At the end of this operation, the result can be expressed in terms of flavor: sour, bitter, salty, sweet (and their combinations) and in terms of taste: of bicarbonate, alkali, metal, chlorine, hydrocarbon, tangerine (which corresponds to the oxidation of traces of hydrocarbons), pharmaceuticals, chlorophenol, earth, mud, sea, mold, or moldy cork (characteristic of herbicides and pesticides).

Consumers of tap water can perform these rough tests for themselves. Among the "physicochemical parameters related to the natural structure of water" is temperature, which must not exceed 77°F [25°C]. However, measurement of the "hydrogen-ion activity" (pH) is more complicated, related as it is to the concentration of protons in moles per liter. The pH is a measure of acidity, which increases with decreasing pH. The pH must be "greater than or equal to 6.5 and less than or equal to 9." The maximal allowable concentrations of ions are as follows: chlorine (250 mg/L), sulfates (250 mg/L), magnesium (50 mg/L), sodium (150 mg/L), potassium (12 mg/L) and aluminum (0.2 mg/L). These are even more difficult to monitor, although the dry extract can be used, which must not exceed 1500 mg/L at 356°F [180°C]. The cations and anions just mentioned are found in almost all normal water. Their presence is fortunate, moreover, because as anyone who has ever had the unpleasant experience of tasting distilled water will report, it is undrinkable. The presence of ions is thus necessary for water to have the appropriate organoleptic properties. Curiously, the decree does not explicitly mention a weight limit for calcium content, which forms incrustations in our pipes, except as skewed by the "alkilinimetric titer," which measures what we call the "hardness of water." In France, this parameter must not exceed the national measurement of 50 degrees. Many waters contain calcium, and people who like it claim that the most pleasant water to drink is rich in calcium; more-

27

over, it has the advantage of rinsing well. Water that is low in calcium makes laundry and dishes difficult to rinse thoroughly.

Several substances are considered *undesirable*, beginning with nitrates, whose level must not exceed 50 mg/L, and other nitrogenous substances such as nitrites, ammonium, and nitrogen measurable by the Kjeldahl method. Hydrogen sulfide and phenols are evaluated by the sense of smell, which is extremely sensitive to the odor of rotten eggs. The level of hydrocarbons (that can be extracted with carbon tetrachloride) is limited to 10 micrograms per liter (μg/L). The law has also set limits for the content of iron, manganese, copper, zinc, phosphorus, fluorine, and silver.

Even worse than undesirable, some substances are considered *toxic*. The list is disturbing. Arsenic, cadmium, cyanides, total chromium, and mercury must not exceed 1 μg/L, nickel and lead 50 μg/L, antimony and selenium 10 μg/L. Then there are the polycyclic aromatic hydrocarbons (PAH). Naturally, the French decree does not mention the entire periodic table; only some of the many molecules and ions that can be found in water are mentioned.

After minerals, the live organisms, or what are known as *microbiological parameters*, are discussed. In this case, the volume of water in which it is acceptable to find *a single* pathogenic organism is given. Obviously, if an enormous volume were considered, it would have to contain some! Not a single one of the now infamous salmonella organisms is tolerated in 5 liters of water, no pathogenic staphylococci in 100 mL, no pathogenic enterovirus in 10 liters. The list goes on: no heat-tolerant coliform bacteria or fecal streptococci in 100 mL, not more than one sulfite-reducing anaerobic bacterial spore in 20 mL, and so on.

Last but not least are the newest arrivals in the toxicity parade, the waste products of modern civilization —*pesticides* and *related products*, the individual levels of which must not exceed 0.1 µg/L for insecticides, fungicides, or herbicides such as aldrin, dieldrin, and hexachlorobenzene (0.5 µg/L altogether), and not more than 0.5 µg/L for chlorophenols such as PCTs (polychlorotriphenyls) and PCBs (polychlorobiphenyls).

We must add that gases dissolve in water. It is for this reason that oxygen and carbon dioxide from the air are found in water. We know that fish cannot live in river water which lacks oxygen. The presence of carbon dioxide allows water to dissolve the calcium in rocks; that is how concretions such as the stalagmites and stalactites seen in caves are formed. Industrially treated water can contain chlorine, an undesirable gas whose concentration is monitored.

WHY THE GOVERNMENT?

The government monitors the purity of the water we consume and promulgates regulations that determine how and what kind of analyses are performed, and how often they are performed, depending on population density and amount of flow. However, French decrees also specify standards concerning the quality of untreated waters, mostly "natural" waters from rivers and lakes that can be trapped by water purification plants and treated to make them fit for drinking. By comparing some figures, the magnitude of the work to be done can be appreciated! Although the mandatory tests do not mention odor or taste, the color is limited to 200 mg/L of platinum after filtration, nitrates may not exceed 100 mg/L, which seems quite modest, and hydrocarbons cannot be above 1 mg/L. The level of undesirable and toxic substances is set at levels comparable to those expected in the final product, but there is much more tolerance for living organisms, whose levels reach astronomical numbers: 20,000 heat-tolerant coliform bacteria, and 10,000 fecal streptococci per 100 mL of water! However, these tiny beastly creatures are relatively easy to kill.

The government is directly involved in this multiplicity of controls because it is responsible for the public health and also must allay the fears of its citizens who consume a sensitive product that is perceived as an integral physiologic component of the human body. Because it is impossible for water to be totally pure, it is absolutely necessary to avoid any incriminating doubt that would provoke rash behavior or panic. However, the potential sources of pollution are so numerous that it is necessary to be vigilant on all fronts, and to monitor everything. Sophisticated, sensitive technology must be enlisted and the competence of the professionals who use it must be certified to ensure, at every juncture and at all times, as universal and unshakeable a confidence as possible.

The list of substances to be analyzed may be frightening, seeming to take delight in lining up the threats, but we must not forget that many of the metallic elements mentioned play an important—if not essential—role in our biological make-up. Thus it is necessary to consume small amounts of these elements. Selenium, for example, is reputed to have a protective effect in preventing cancer. But other substances are undeniably harmful. Lead is suspected of having contributed to the downfall of the Roman Empire by destroying the minds of the elite who used lead cooking utensils. Since water pipes are often old, it is occasionally observed that in some places the lead level greatly exceeds the standard (as high as $520 \, \mu g/L$ in some cities!). This phenomenon can be readily explained. Water having a low calcium content is not very hard, but is acidic because it contains carbon dioxide, which has a tendency to dissolve the lead carbonates that form over time in pipes. These carbonates are analogous to the white lead that was formerly used in paint and which causes severe poisoning in children who ingest particles of paint from old walls. The solution to this problem, which can pose a serious public health threat, consists in replacing these fixtures with plastic pipes. Aluminum, included above in the category of natural physicochemical parameters, is the subject of controversy. It has recently been suspected of contributing to Alzheimer's disease, so much so that countries such as the United States have set a maximum concentration that is lower (0.05 mg/L) than the traditionally accepted standard. However, since aluminum is a natural component of clay, not to mention the fact that water is ultimately used with cooking utensils, regulating the intake is probably difficult to enforce.

Water is a medium which, depending on certain conditions, accepts or rejects chemical substances. That is one of its properties that cannot be circumvented: water always contains *something*. This can be all the better appreciated as the development of chemical analysis techniques now makes it possible to detect ever smaller amounts of elements or molecules. Technology is thus constantly lowering the limit of detection, while, at the same time, rapid routine analytic methods are multiplying. For almost every chemical we can today assign a figure expressing its concentration. The public is not always aware of the order of magnitude of the units of measurement associated with such figures, that is, the scale on which they are measured. The new analytical methods have such fine thresholds of detection that even extremely small amounts can be measured. But the mere indication that something is present, in whatever minute amount, suffices to trigger fear in a consumer who is not familiar with the actual numerical relationship of the molecular populations involved (one "ppb," or part per billion, for instance, corresponds to *one* member of the entire population of China).

RETURN TO ROMANCE

The French standard, which closely follows the European directive of July 15, 1980, provides for sixty-two parameters for drinking water, forty of which correspond to maximum allowable limits imposed for concentrations. This is a complex matter, for we are drinking a product that is theoretically well-monitored, but the variables of which are so numerous that the safety intended depends on the frequency and extent of the monitoring, since pollution can be fleeting, appear suddenly, then rapidly recede. The satisfaction of the water-drinker, who has his or her own tastes, can never be guaranteed. Thus, the average American does not seem to care much for the ozone-treated water available in France, because the reassuring taste of sodium hypochlorite (chlorine bleach) is missing, while some Africans consider the well water of the Peace Corps volunteers to be too "pale" compared to the water in the watering holes where the local fauna wades, which apparently has the advantage of giving it a special taste! In terms of water, the physiologic senses reserve the right to make the final judgment.

There is an abundance of residual "non-water" that is always found in water, even the most carefully treated to please us. This repudiates the absolute notion of an ideal pure water in a collective imagination haunted by it—a specter manipulated by advertisers in posters and images on our television screens or on the glossy paper of tourist brochures. Water splashing over a young body with thousands of scintillating droplets is taken as a symbol of joy, freedom, and the pleasure of consuming a "real" product." We see no hard water deposits—a classic villain for advertisers—, no nitrates or traces of PCBs. The presence of chemicals, several of them harmless and which can even be "natural," is perceived as a brutal aggression of Industry against Nature, like an unpardonable rape.

And yet, "ideal water" which is suitable for the body is no different from the water defined in great detail in the list of varied parameters for the standards in effect today! It is somewhat of a communications tragedy that Science, Industry and the *Official Gazette* disenchant by a litany of numbers an object that is potentially perfect, whose praises are sung by the poet beloved by the muses! However, we must reconcile ourselves to the fact that, as a *physical* object, pure water does not exist, especially not in Nature!

II

THE MECHANICS

OF WATER

The most obvious property of water is its ability to flow, to escape, to slide through cracks, to slip through the fingers. Another obvious property is that water easily fills its containers—from a bottle to a river bed. However, water knows no limits: torrents and floods confirm its universal propensity to spread in the space available to it until, why not? it covers the entire Earth. A fluid, agile, and available substance, water is also an amazing force which mass repels. Capable of uprooting and carrying off the most solid structures, however, it is immediately impressive in its mechanical properties, the distances it travels, the nature of its movement, the work it can do. Water has played a role in many decisive respects in the birth and development of physics because it has provided several of the first objects of quantitative study in statics and then in mechanics. Rivers, rain, canals, deep wells, and the deafening roar of waterfalls come naturally to mind. Water's confined calm expanses, its peaceful or turbulent flow, its provocative drops, and its rise in siphons invite us to apply the power of calculation to the engineer's art. There is no doubt about the social usefulness of the quantification of observations it has made possible.

The physical and mechanical properties of water are of tremendous importance to humanity. Used for our benefit since ancient times, these properties are still the subject of research carried out with the most sophisticated means, often for the purpose of applications that contribute to shaping the life of our society.

CONTINUITY

Statics and mechanics are sciences of the continuous. They consider objects for which all the abstractly defined *points* are equal, indiscernible, capable of being understood intellectually as indefinitely contractable portions of space, but which nevertheless are so close to one another that no solution of continuity can occur. At the dawn of mechanics, and even much later, while chemistry was already in its prime, there were many scholars who did not accept the fact that matter could be a "collection of massive points *separated* by a vacuum." This idea appeared to contradict ordinary common sense, that of the hand that perceives the hardness, softness, or fluidity of substances.

Water supplied the basic model for a fluid: a flexible, moving body without rigidity that is *continuous*. The limits of the space occupied by a mass of liquid appear to be uncertain, because we can with little effort modify them. We know that there are two types of fluids: liquids and gases. Water vapor is water in gaseous form. Liquid requires a container, it is confined in a shape, whereas a gas occupies all the available volume, it is *expandable*. Gas and liquid can flow, circulating from one place to another. To understand the mechanism of these shifts, liquid is easier to observe and to control. On the other hand, although gas is easy to compress, it is difficult to compress liquid.

It is necessary to accept the fact that a fluid mass is the subject of a local description. The entire volume of a fluid can actually have different properties at different points: differences in temperature, flow rate, and pressure can be observed. To realize this, just plunge your hand into a flowing brook. A fluid thus often appears heterogeneous if its entire volume is considered. To apply mathematics to such a situation, it is first necessary to construct a formal abstraction: decompose the volume to be considered into elements of any form, but within which one can consider a *homogeneous* material, that is, having the same physical properties (pressure, temperature, density, etc.) at each of these abstract points whose juxtaposed sum constitutes a continuous space. Such thought objects are what we call *particles.* In a fluid these particles move with respect to each other, and a mathematical analysis can be made of this dynamic by using a geometric decomposition of the motion of each particle with respect to the others. The rate of translation, rate of rotation and rate of deformation during movement are studied.

The dimension of the particles is highly variable. It is determined as a function of the problem under consideration, but is always large with respect to strict molecular dimensions. In meteorology, such a particle can attain several miles. In general, however, we consider particles to be "small." We accept that they are homogeneous and, moreover, that they are preserved, namely, that they have a behavior that is, after all, *individual* and unique. Naturally, it would be much easier today to see in the molecule the ultimate particle, but *fluid mechanics*, an engineering tool, was developed well before the concept of the molecule emerged. Although it uses rather simple calculation methods, this science quantitatively describes many physical effects (mass and heat transfer, frictional effects). Today we know that the forces actually involved in these processes originate in the individual properties of molecules. Specifically, molecular dynamics, a recent innovation, is based on the calculation capacity of computers. Impressive as it is, this capacity is still too weak to allow concrete questions to be dealt with on the basis of a group of particles visualized on the actual scale of intermolecular distances!

THE MATHEMATICS OF FLOW

Water is contained in a recipient. Whether artificial or natural, it has walls: a bottom and sides. Water is thus in contact with the solid material that forms these walls. It also has a free surface, which is flat if the only force exerted on the particles of the volume considered is that of gravity. Regardless of how free it is, the surface of water is also an area of contact between the liquid phase and the gaseous phase (air). Water in a container collides with two physical boundaries: between liquid and solid and between liquid and gas.

The particles within a fluid like water "rub" against each other when they move. When they come into contact, they have a tendency to carry each other along, providing, of course, that an external potential causes the collection of particles to move. This friction due to contact—or the effect of adherence—constitutes the *viscosity* of the fluid. Regardless of the popular usage of the word, real fluids are always viscous. In terms of mathematical abstraction, however, we can consider ideal, nonviscous fluids. This abstraction allows physicists to ignore the weak forces of viscosity so that, for example, to describe flow they imagine entire zones of *free fluid* in which the trajectories of particles are considered to be parallel. But near the walls or obstacles arising in the middle of a current it is necessary to consider this viscosity, as well as the phenomenon of adherence of the fluid to the walls, and to take into account the braking effect applied as a result of the collision of moving particles with fixed solid masses.The rate of motion of the particles is modified by contact with the walls over a rather large distance across a zone known as the *boundary layer*.

The description of the flow of water in a pipe was one of the first objectives of the science of hydraulics. By taking into account the respective spatial requirement ratios between the boundary layer and the free fluid part, starting at the mouth of the pipe, the conditions leading to a stable rate of flow can be defined. For example, these techniques can be used to describe a slow laminar flow with a distribution of speeds through a cylindrical cross-section of the pipe that follows a parabolic law. In this case, the flow of particles is more rapid along the axis of the pipe because the zone of free fluid has been swallowed up by the enlargement in the size of the boundary layer.

Once the behavior of particles of water in a pipe is known, it becomes possible to derive laws of fluid dynamics in the case of flows known as stationary (or permanent), to indicate that the data provided by the measuring apparatus on the physical conditions of the flow, such as pressure, rate, temperature, and density, are independent of time. The industrial techniques used for the movement of fluids and the transportation of energy all use this type of flow. Equations are sought to link the different physical quantities at each point of the pipe. To obtain these equations, an idea related to infinitesimally small quantities in mathematics is used. A thought object known as a *stream tube* is isolated, a small ideal tube of fluid that combines the lines of flow, giving material form to the trajectory of an infinite number of particles. A good illustration of these stream tubes can be given using a pipe with a tiny diameter to inject a dye into a calm stream of moving water: the dye particles "rub" against the water particles and follow them in their movement, thus making them visible. The visualization of stream tubes is not a mere pastime. It is an important technical problem whose solution requires all sorts of clever tricks with models: air bubbles or polystyrene beads, unctuous colored liquids, etc. Several hydrodynamic phenomena were discovered by using these methods; in particular, it became possible to study the conditions of separation from the walls by laminar boundary layers, which cause swirling zones such as wakes to appear.

If everything is stationary, two sections of the stream tube, possibly with different surfaces, must allow the same amount (weight) of fluid to pass within the same amount of time. Obviously, then, as the cross-section becomes smaller, the rate increases. This explains, mathematically, why the flow of rivers increases when they are suddenly narrowed by some geographical obstacle.

A mass of moving water carries energy. To study flow that occurs solely as the result of gravity, the law of conservation of energy is applied. Using the model of a line of flow, the equation relating energy, pressure, and density of the fluid can be established. This equation was formulated in 1738 by Daniel Bernoulli (1700–1782). In the ideal case, energy can be defined as the sum of potential energy and kinetic energy (which depends on the square of the velocity). For a horizontal tube (constant energy), it can be deduced from this equation that the pressure is inversely related to the velocity and therefore directly proportional to the cross-section of the tube. This result does seem to be somewhat counterintuitive because, conversely, the pressure is lowest when the velocity is highest! This phenomenon is used in the construction of devices that measure water flow, by comparing the pressures in a large cross-section of a sewer-system duct and in a narrower derivative pipe. Another application, which is far from frivolous, consists in blowing air across the highly tapered end of a glass tube, thereby creating a vacuum at the opening, which, when applied to the end of another small tube immersed in a reservoir, allows liquid to be aspirated which is then sprayed by an air jet. This is the principle of the atomizer.

Naturally, the Bernoulli equation is perfectly suited only in the case of ideal mathematics. In fact, in real water mains, there are losses of mechanical energy in the form of heat from losses due to friction. This loss of heat is greater when the ducts are longer and more tortuous. We should add that the permanent flow condition imposed by the equation is never more than partially satisfied. Most of the time, the values of the measurable physical quantities fluctuate considerably with time about the mean. If the fluctuations become too great, then we have what is called *turbulent flow*: the stream tubes mix together, lose their individuality, and the range of velocity distribution becomes narrower. The simplicity of the old historical equations no longer describe the situation, and the detailed description of flow becomes a problem to which the power of sophisticated computers is applied. Using these techniques, models are established, creating false color maps on which different hues represent the values of different parameters, such as velocity. These maps can be used to study images of turbulence, which are often quite beautiful.

For the engineer, however, the difficulty involved in the calculations can be avoided rather easily. In hydraulics, dimensionless numbers are used which relate measurable quantities, such as the diameter of the pipes, the average flow rate, the density, and the viscosity of the fluid, to each other. Such numbers—like Reynolds, Froude, or Mach numbers—allow us to know whether or not a flow pattern will be laminar, turbulent, or intermediate. Thus, the flow of water through a cylindrical pipe will be turbulent if the Reynolds number has a value in the thousands. The advantage of such dimensionless numbers is that the result obtained has a value that is independent of the experimental conditions. In this way, work on experimental models or reduced-size models retains the same numbers associated with actual flow.

The fundamental phenomena of real flows such as turbulence, separation, and vorticity can be studied by using both theoretical models and numerical approximations within the framework of a three-dimensional approach. Viscosity effects, boundary layers, turbulence and approximations of the perfect fluid constitute the principal ingredients of the problems considered. This is the case, for example, with the problem of the pressure distribution about a submarine hull. In practice, a grid technique is often used, which consists in carrying out numeric calculations only for points that are rather close together on a grid and extrapolating the values for the space between the intersections. Despite this choice, which limits the multiplicity of calculations, the most powerful computers available (such as Cray computers) are still necessary for evaluating the mechanical constraints to which high technology constructions that must move through water or are buffeted by the flow of water are subjected. Bridges, dams, and dikes are the privileged and spectacular applications of these calculations. Today, the available power of calculation and the progress in establishing theoretical models lessen the necessity for experimental models that once served to deduce by analog methods the mechanical constraints exerted on civil engineering structures and watercraft. Nevertheless, such experimentation continues to be extremely useful for discovering general properties, which must then be encoded by calculation, or for locating unsuspected weak spots. Problems of walls and wakes which are always specific to the particular geometry of the object under consideration often require the use of experimental models in hydrodynamic tunnels to aid in visualization of the problem.

In the case of ships, the agitation of the surface induced by the force of the wind presents another problem that can be solved by using these same techniques. If the agitation is more or less periodic, swells are involved, which cause oscillations that may destabilize a floating body subjected to stresses that are variable and difficult to model. The dimension of time and the random nature of meteorology are thus added to the problem of stationary hydrodynamics. We know that objects having a certain suitable density will float due to the effects of the famous principle of Archimedes: an object immersed in a liquid is subjected to an upward force applied to the center of the mass of displaced fluid which is equal to its weight. However, this respected principle tells us nothing about the dynamics of movement associated with a rough sea!

THE SKIN OF WATER

The concept of the boundary layer allows us to take into account the slowing effect exerted on the displacement of molecules when water comes into contact with a solid immobile surface. We have seen that the contact which occurs at the surface of a layer of water is similar to that occurring with a gas, most often air. In the case of an extensive horizontal plane, the surface is flat. But if the separation between the water and the air is observed in a narrow glass tube, it can be seen that the surface of separation, the "meniscus," is concave, as though the water were climbing the walls of the tube. To understand such phenomena that can be observed at the surface of water,

45

we must once again rely on mathematical thinking and use our imaginations to cut a narrow strip of liquid at the air-water interface, as if we were removing a thin film of plastic from the surface. If we were then to break this film by pulling horizontally at the edges, it would be necessary to develop a certain force, known as *surface tension*. Its numerical value is modified by chemical agents that can be present at the air-water boundary as a function of their density. Among these chemical agents, surfactants render the superficial film elastic: it can be pulled and stretched a great deal before it breaks. We can readily form a concrete image of this film by observing the diaphanous envelope of a soap bubble. Indeed, in this particular case, the water film that is sandwiched between two layers of perfectly aligned oleic acid molecules is stretched. For an even more spectacular demonstration, thin iridescent surfaces can be made by dipping a wire loop into a soap solution. Beautiful colors are observed if the thickness of the film does not exceed 10 to 20 μm. Surface tension involves intermolecular mechanical forces that tend to reduce the size of the surfaces. If these forces are dominant (which occurs with low volumes of liquid), the well-known form of the more or less spherical *drop* ensues, which minimizes the surface-to-volume ratio.

We can imagine that the same surface film exists for water that is in contact with the surface of a solid. The nature of the relationship between the water molecules of the film and the first few atomic layers of the solid will vary according to the chemical nature of the solid. There are two possible cases. In the first, the water molecules adhere or stick to the surface; they then have a tendency to spread and, relative to a horizontal plane of water, to "climb" the vertical walls. This is what happens with glass; the meniscus is concave, and it is said that the water "wets" the glass. In the other case, however, if the elementary components of the liquid and those of the solid "repel" each other, the geometry of contact is reversed: the liquid, instead of spreading, "curls back" on itself in a drop. Thus, we can observe that the meniscus of mercury in a small glass tube is convex, since mercury does not wet the glass. Nor does water wet impermeable surfaces, such as rubber: it stays in drops and runs. These effects of the adhesion of liquids to surfaces and to the walls of containers, specific for each type of liquid-solid pair, are combined in the general term *capillarity*. This is a superficial phenomenon that may become quite significant in a mass volume if the surface-to-volume ratio is very large. Such is the case for many natural systems, especially plants, which absorb water from the soil through the phenomenon of capillarity. Similarly, we can explain the nature of porous materials, which are solids having cavities that are wetted by water, which can slide and pass through them. There are relatively new fabrics (such as Gore-Tex) that are permeable to air, thus pleasant to wear, but which are not wetted by water and have the additional advantage of being waterproof. Likewise, water *percolates* rather easily through clusters of granular solids. These are all practical questions previously

47

disdained by theoreticians but that can now be understood by means of our new conceptual tools which provide a "geographic" description of microsurfaces on a molecular scale. In the area of capillarity, there are certain unusual cases: some molten salts, for example, "climb" up the walls of the crucible in which they are being melted, virtually to the point of spilling outside the crucible. The forces associated with these surface effects cause water to rise, in relation to a horizontal level, in what are aptly called capillary tubes having internal diameters of only a few millimeters. These tubes are used to measure surface tension. The phenomena of surface tension and capillarity are of huge economic importance—from household detergents to high-tech agricultural techniques.

GAS AND BUBBLES

Water molecules that wander about on a free surface find themselves in contact with gases in the air. Exchanges occur between water and gases: some molecules of water escape into the air while others coming from the air are deposited on the surface and mix with the mass of water. If the number of molecules of water leaving the surface is equal to the number of molecules of water that are added to the water, it is said that the liquid and gas are in *equilibrium*. The gas exerts a certain amount of pressure on the liquid, and the molecules of water in the gas contribute to a fraction of this pressure. At equilibrium, the partial pressure of the water in the gas defines the *vapor pressure*, which depends on the temperature: the more the water is heated, the greater is the agitation of the molecules, and the more easily they escape from the surface: As their number in the gaseous phase increases, so does the vapor pressure. This phenomenon is universal. As strange as it may seem, even ice has a vapor pressure; it is low, but not negligible, and can be put to industrial use. A solid that loses its material in the form of vapor is said to "sublime." When ice is sublimed under vacuum, water escapes in the form of water vapor. This is the basis of the process of lyophilization, better known to drinkers of instant coffee as freeze-drying. When the partial pressure of the water in the air is lower than the equilibrium vapor pressure (known as saturation) at the temperature of the sample, the water will evaporate. If the partial pressure of the water in the air is greater than the equilibrium vapor pressure, then the water will condense in the form of fine water droplets on dry surfaces. This is what happens when dew forms.

We can thus see how exchanges of material at surfaces are constantly occurring. The total passage of liquid to vapor (or solid) form represents a change of state, a phase transition, which occurs at a certain temperature and at a certain pressure. Pressure and temperature are said to be *state variables*. For each fluid there is a *critical point* characterized by precise numerical values for pressure and temperature, beyond which there are no longer any differences in pressure or temperature between liquid and gas: this is the *critical fluid* zone.

If a certain amount of water at a given pressure is brought rapidly to the temperature where the stable phase is the vapor, instead of simply evaporating from the surface, the water will begin to boil. Why are bubbles produced? They indicate the first appearance of a new phase in the medium that was previously homogeneous (known as *nucleation*). This phenomenon is thermodynamically inevitable, and then all the beautiful equations of mechanics disregard the very principles on which they were founded. They become applicable again only when all the water is transformed into vapor, hence into a fluid which is now compressible and which can do "work." We know that the steam engine is a means of transforming heat into mechanical work.

In all situations where the saturating vapor pressure is reached within a liquid and where bubbles are formed, classical mathematics are no longer rationally applicable to the process under consideration. This is what occurs in the upper portions of siphons, and more generally in certain mechanical devices where a decrease in pressure occurs as a result of an increase in the rate of fluid movement, as is the case with pumps, ship propellers, and nozzles. The appearance of bubbles then induces the phenomenon known as *cavitation*. Engineers must take great pains to avoid cavitation, because parts of the pipes —and, more generally, their exposed solid surfaces in affected areas—are subjected to intense vibrations and mechanical erosion, due in particular to implosions of the bubbles. This generates shock waves that locally subject the material to brief and irregular changes in temperature and pressure which may, in the long run, exceed the overall mechanical strength limits and cause large-scale breaks.

THE CONSTANTS

For a long time, we have known most of the quantities involved in the applications of fluid mechanics and thermodynamics to water. The numerical values were determined by scholars of the 18th and 19th centuries with the instruments they invented and which are still in use today. It is not necessary to know the intimate molecular structure of water nor even its chemical composition to collect and use these data. Density, surface tension, viscosity, vapor pressure, temperatures of state changes, specific heat (the energy which must be supplied to increase the temperature of 1 g of water by 1°C), thermal conductivity, which depends on diffusion (the movement of particles that transmits the heat), and the speed of sound, which is related to the existence of a certain compressibility of water, are all parameters that can be precisely measured within a wide temperature and/or pressure range and are included in detailed tables known as *tables of constants*.

51

All this information forms a rich body of experimental data, but one that remains empirical because the scientific explanation for the specific numerical values found is possible only through simulation calculations, which assumes a considerable theoretical effort that itself requires a precise knowledge of the molecular laws on which the structure of water is based. This effort is all the more indispensable because if we compare water with other liquids it appears that the measured values are well outside the average limits, varying with temperature in an astonishing fashion, which at first sight seems incoherent. From this comparison emerges the idea that water is an unusual or, as some would say, an "abnormal" liquid.

This is conceivable if we consider a certain number of remarkable and "strange" things that affect the major physical properties of water. For example, density reaches a maximum at 3.982°C; the coefficient of compressibility exhibits a minimum at around 45°C, which results in a maximum rate of propagation of sound waves at 70°C, and decreases with increasing pressure; viscosity, very high in absolute value, drops to a minimum when the pressure increases, with specific heat also evidencing a pronounced minimum at 37.5°C; surface tension and the dielectric constant (another important physical property), both very high, decrease with increasing temperature. What is disturbing about all these measurements, aside from the exceptional numerical values, is the absence of the reassuring linearity that allows reasonable extrapolations, as well as the range of temperatures at which the variation in a physical constant changes direction. Later on, we will see the profound reasons for these oddities.

III

ANALYSIS OF WATER

CHEMICAL COMPOSITION

The invention of the steam engine was one of the first industrial applications of the "motive power of fire," in the words of Sadi Carnot (1796–1832). The principle is simple enough: the energy derived from combustion can be used when it is stored in the form of heat in water, and can then be transformed into mechanical movement as a result of the passage of water from the liquid state to the vapor state. This combination of fire and water in the same machine changed the face of the world. It was the need to pump water out of mine pits that was the driving force for research that began at the beginning of the 17th century. Between 1690 and 1710, Denis Papin (1647–1712), Thomas Savery (1650–1715), then the British mechanical engineer Thomas Newcomen (1663–1729) invented machines that would actually be used and whose operation was in all cases based on the sudden condensation of water vapor: the vacuum thus produced forced the aspiration that caused the water in the well to rise. These primitive monsters, which consumed a great deal of energy, were used especially in England, and also in Holland to drain land. While looking for a way to economize on coal, James Watt (1736–1819), the son of a Scottish laborer, seized upon the idea of taking advantage of the latent heat contained in steam. (The amount of energy required to vaporize water is considerable—539.55 cal/g at 100°C.) In 1763, by improving the engineering, using heat insulation to prevent unnecessary cooling, and dealing logically with condensation, Watt developed a modern prototype. The dual-purpose machine, with its flywheel and governor, appeared in factories in 1781; it was to permit the launching of the worldwide industrial revolution.

In addition to inventing the steam engine, James Watt was also a chemist and just barely missed adding a supplementary title to his glory. At that time in England, France, and Germany, chemists were interested in determining the composition of air and understanding the principle of combustion. There were arguments about the existence of "phlogiston," a subtle fluid thought to be released during combustion; the reasons for the ability of certain gases to burn and the inability of others to burn was not understood. In particular, this "inflammable air" envisioned by the alchemist Paracelsus (1493–1541) was studied, and the techniques for obtaining inflammable air were perfected in 1765 by Lord Henry Cavendish (1731–1810). He reacted vitriol (sulfuric acid) with iron, which produces what is known today as hydrogen. Later on, in 1774, by slowly heating mercury in a closed chamber to obtain the oxide, then decomposing it in a bell jar with solar energy and a glass lens, another English scholar, Joseph Priestley (1733–1804), discovered oxygen, "dephlogisticated air," which does not burn but acts on everything that can.

James Watt belonged to the Lunar Society, a group of scholars who met only on nights when there was a full moon and kept up a voluminous correspondence to keep each other apprised of their respective experiments. In 1776, Pierre Joseph Macquer (1718–1784) in France and Waltire in England observed the formation of drops of water on a "white porcelain saucer" during the combustion of "inflammable air." Then, in 1781, Waltire noticed that water forms when electric sparks are caused to burst in a mixture of this dry inflammable air and ordinary dry air. This observation was repeated by Cavendish and by Priestley, two years later, with dephlogisticated air.

Watt had the intuitive feeling—and rightly so— that water is not a simple substance, but is the result of the transformation of these gases during combustion. He verified this hypothesis on the basis of weight, and on April 21, 1783, prepared a letter to the Royal Society, which he was to retract a few days later. In the interim, he had received a letter from Priestley (dated April 29, 1783) from which he learned that his own experiments, which he believed demonstrated the transformation of water into ordinary air, had been skewed by the porosity of the porcelain used for the tubes and retorts; when heated, they allowed air from the outside to enter into the experimental chamber. Priestley thought that this sad observation also invalidated the "beautiful hypothesis" of his friend. We know, from a letter addressed to Priestley on May 2, 1783, that Watt nevertheless maintained his opinion, but prudently withdrew his letter to the Royal Society. It would not be read until April 22, 1784, too late for him to claim the credit for one of the major discoveries of modern science.

During this period, French scholars were working on the same question. On June 24, 1783, in the presence of the secretary of the Royal Society of London, Antoine Laurent de Lavoisier (1743–1794) and Pierre Simon Laplace (1749–1827), using a carefully-constructed glass apparatus whose airtightness and glass quality were carefully checked, achieved the *synthesis* of 19.17 g of water "as pure as distilled water" by burning hydrogen in oxygen. This experiment showed that water is 85% *oxygen* and 15% *hydrogen* by weight. The two elements were later named by Lavoisier in 1787: "that which produces water" he called *hydro-gen* and "that which produces acids" he named *oxy-gen*. The *decomposition* of water was achieved by Jean-Baptiste Meusnier (1754–1793), a student of the mathematician Gaspard Monge (1746–1818), by passing water vapor over iron and charcoal heated to glowing red. The results of this experiment, which confirmed the synthesis of water, were announced to the French Academy of Sciences on April 21, 1784. Meanwhile, stimulated by the news from France, Cavendish, braving the snow and cold, described his own work before the Royal Society on January 15, 1784.

Lavoisier declared triumphantly in his *Traité élé-mentaire de chimie* [Elementary treatise of chemistry], published in 1789: "Water is no longer an element for us.... Water is not at all a simple substance...water is a compound."

To complete the knowledge of the chemical composition of water, it was necessary to perform another great experiment that only became possible on March 20, 1800, when the Italian physicist Alessandro Volta (1745–1827) wrote to the Royal Society announcing his discovery of the electric battery. This was the first step in domesticating the elementary particle that would later be known as the electron. This discovery can be considered to be the basis on which our modern technical civilization is founded, because it was the first step in the controlled production of electric current. Volta's letter was not read to the Royal Society until June 26, 1800, but in the interim the English scientists Sir Anthony Carlisle (1768–1840) and William Nicholson (1753–1815), became aware of it. They immediately made a battery with zinc and silver, and on April 2, 1800, used it with some copper wires to perform the first electrolysis of water. By replacing the copper with platinum, which is resistant to oxygen, they obtained the volumetric composition of water: 72 volumes of oxygen to 143 volumes of hydrogen, or a ratio of 1 to 2. The path to the formula for water, H_2O, was now cleared.

ISOTOPIC COMPOSITION

Thus it can be considered that the question of the analysis of water was solved at the beginning of the 19th century. Nevertheless, in modern times, a considerable complication has been added to this analysis. Indeed, when physicists studied the structure of the periodic table, they determined that an atom of an element is composed of a central atomic nucleus, containing a specific number of protons (elementary particles, units of atomic mass having a positive electric charge) and a cloud of a like number of electrons (much lighter elementary particles having a negative electric charge) that orbit around the nucleus at a relatively large distance compared to the dimensions of the atomic system. An element is characterized by the *number* of protons in its nucleus. In the case of a *neutral* atom, the number of protons is balanced by an equal number of electrons. It is the number of electrons that determines the chemical properties, whether or not the number of electrons is equal to the number of protons. But in addition to the protons, the nucleus may also contain neutrons, other elementary particles that have nearly the same mass as protons, but no charge. The number of neutrons in the nucleus of the same element may vary, but of course the chemical properties that depend on the number of electrons do not. An element can thus have several varieties, each having different numbers of neutrons, but since they have the same number of electrons, the chemical properties dependent on electrons are the same; these varieties are known as *isotopes*. Since the atom's mass depends essentially on the mass of the nucleus, the isotopes have different masses. Consequently, the physical properties that depend on mass are different for different isotopes.

Hydrogen, the first element in the periodic table, is composed of one proton and one electron. The isotope of hydrogen, *deuterium*, whose nucleus contains one proton and one neutron, has twice the mass of hydrogen—a considerable change! The American chemist Harold Urey (1893–1981) and his group announced the discovery of deuterium in January 1932. At that time, the analysis of light emitted by discharge tubes containing gas at low pressure was one of the spectrographic methods used to characterize new or poorly identified elements. The spectrum of atomic hydrogen is formed by sharp lines covering a wide range of wavelengths. Measurement of the energies associated with these lines made it possible to discover the mysterious nature of the atom. The origin of the lines lies in the electronic structure, which depends to some extent on the mass of the nucleus. For a given isotope, the same lines may thus appear slightly shifted. Urey used high-resolution spectroscopy to observe the weak satellite lines of the isotope having mass 2 alongside the principal lines in the hydrogen spectrum. W. Bleakney confirmed the existence of deuterium several months later by mass spectroscopy. Shortly thereafter, James Chadwick (1891–1974), Frederic Joliot (1900–

1958), and Irene Curie (1897–1956) identified the neutron as being involved in certain radioactive phenomena.

Another isotope of hydrogen, *tritium*, has a nucleus composed of one proton and two neutrons. This isotope is not stable; it is radioactive. It disintegrates by emitting an electron (beta radiation). Its half-life (the amount of time it takes for one half of the initial number of atoms to decay) is 12.26 years. Tritium is produced during different types of nuclear reactions, such as the collision of two atoms of deuterium, or the collision of a neutron and a deuterium atom. Tritium is involved in the explosive mechanism of the hydrogen bomb, in which a deuterium atom collides with a tritium atom to produce a helium atom (second element in the periodic table) and a neutron, releasing an enormous amount of energy. In the atmosphere, tritium is produced by the action of cosmic radiation on nitrogen: its infinitesimally small concentration in rain water corresponds to the ratio $^3H/^1H = 10^{-18}$.

Oxygen, whose nucleus contains eight protons, has three stable isotopes that make up the population of natural oxygen. The most abundant isotope, having a mass of 16, represents 99.758%, the isotope of mass 17 represents 0.0374%, and the mass 18 isotope represents 0.2039%. There are also unstable artificial varieties having masses of 14, 15, and 19. Considering that one of the components of the water molecule has two natural isotopes and the other has three, we can see that there are many possible varieties of this molecule in addition to H_2O. The most significant of these are *heavy water*, known to chemists as deuterium oxide (D_2O), and deuterium hydroxide (HOD). All the water on our Earth contains hydrogen and deuterium in the ratio 6400/1. This means that for every 6400 hydrogen atoms there is also *one* deuterium atom. On the other hand, the $^{16}O/^{18}O$ ratio is 489.3/1. Of the 18 possible combinations, the most common are:

$$^1H_2\ ^{16}O/^1H_2\ ^{18}O/^1H_2\ ^{17}O/^1HD\ ^{16}O, \ldots = 97280/2000/400/320, \ldots$$

From 15 liters of ordinary water, 1 g of heavy water (D_2O) can be extracted by taking advantage of its physical properties: density 1.107 at 25 °C (with a maximum at 11.4 °C, and not at 4 °C as is the case with H_2O), melting point 3.813 °C, boiling point 101.4 °C, which allows it to be prepared by fractional distillation techniques. Since it is preferable to carry out electrolytic decomposition using light water, heavy water can be concentrated this way, as has been done in Norwegian factories since 1940. The properties of heavy water are used during the process of nuclear energy production to "slow down" the neutrons. The energy they carry is actually spent in their elastic collisions with the nuclei of deuterium atoms, because—unlike the case with hydrogen, which can easily capture a neutron to form deuterium—it is difficult for deuterium atoms to capture neutrons to form tritium.

The existence of different isotopes of water has made it possible to develop the basic techniques for nuclear energy currently available. The "battle for heavy water" was a seminal event having both military and strategic implications. This event exerted a decisive influence on the course of World War II; here we can see a precise example of the role scientific knowledge plays in the genesis of power in the modern world. Today, similar molecules fuel our fears, but for quite different reasons. Because of radioactive isotopes, the history of the Earth has been preserved: we can reconstruct the snows of yesteryear, gain awareness of the evolution of climates over the past several thousand years, and appreciate the capriciousness of the episodes of rain and good weather. In this area, nothing really seems stable for very long; fluctuations are the rule rather than the exception.

Starting with these data, researchers can thus speculate on the causes of natural climatic variations and attempt to clarify the effect of the contribution of our industrial activities on the evolution of phenomena that are still poorly understood. Actually, mass spectrographic analysis allows us to measure with considerable precision the $^{16}O/^{18}O$ ratio of water, ice, or snow. Glaciologists use this method to reconstruct the climates of the past. The molecules of water containing heavy isotopes evaporate less easily than lighter water molecules from the surface of the ocean during cold weather, so the atmosphere is enriched with ^{16}O, which is found in the annual snowfall preserved in the form of ice layers piled up in the inland ice at the Earth's poles. The isotopes in water thus allow us to decode the climates of the past by reconstructing the average annual temperatures, for which wide and sudden variations were found to have occurred. In this way it was discovered that 10,700 years ago, the average temperature in Greenland increased by 7°C over a period of 20 years.

IV

THE ISOLATED

WATER MOLECULE

Data on the chemical composition of water has resulted in the formula for the smallest elementary entity: H_2O. It seems entirely natural to us today that this building block is a *molecule* formed by *two atoms* of hydrogen and *one atom* of oxygen. Although the formula expresses the minimum number of atoms that participate in the formation of this molecule, the mere statement of these numbers says nothing about the way in which these atoms are arranged in space. A long time passed before anyone even admitted that this question should be asked. *Stereochemistry*, developed around 1874, was the subject of much criticism on the part of the chemists of the time, such as Marcelin Berthelot (1827–1907). True enough, it implied acceptance of the existence of atoms. At the end of the 18th century, Abbé René-Just Haüy (1743–1822) had already been the object of severe criticism by his powerful contemporaries, including Claude Berthollet (1748–1822), for proposing the idea of "integrated molecules" to explain the regular geometric forms of crystals. Particularly shocking was the kind of rigid image attributed to molecular structure. Not only did the molecule appear to be imprisoned in numerical ratios, but it was discovered to be trapped in a place frozen in space. Many found this ridiculous, preferring to imagine swirling particles, dancing partners, the freedom of vibrating, twisting and rotating. They invented a sort of chemistry constructed on the model of a ballet whose characters sometimes vigorously embrace and at other times nonchalantly change partners, touching each other only by their fingertips. At the time, this dynamic vision seemed to conform better to the vigorous exchange of matter in chemical reactions.

Chemical notation was first based on empirical formulas, such as H_2O. Then, especially for the needs of organic chemistry, ever greater use was made of structural formulas (such as H–O–H), in which approximations for the way bonds can occupy the space around an atom are drawn as flat or three-dimensional representations.

Graphical representations incorporating several basic principles have proven to be quite useful in facilitating thinking by allowing new reactions to be predicted: new products can be imagined by substituting one element for another, for example, by replacing hydrogen with chlorine. The classical forms of chemistry symbols are still used today because they are well-suited to the presentation of scientific results in printed form. This use of symbols greatly simplifies discourse. However, the reality of molecular structure is more complex. We know this to be true because we have learned to use matter-radiation interactions to obtain precise information, first on the structures of crystals, then on the structures of individual molecules.

These technologies have expanded explosively over the last twenty years as a result of progress in vacuum technology, electronics, and the use of computers. Basically, equipment incorporating three elements is used. First, a source of radiation is needed corresponding to either some part of the photon electromagnetic spectrum—from gamma rays to radio waves—or to a beam of elementary particles, such as electrons or neutrons. Electromagnetic radiation may be coherent (as is the case with lasers), or not coherent, and it may be very intense (for example, synchrotron radiation). Then the material to be examined and the beam are made to interfere, the components being scattered, absorbed, diffused, diffracted, or reflected. Finally, detectors analyze the position, the energy, and/or the intensity of the radiation after interaction, for example, by using a photographic plate. A mathematical interaction model allows the researchers to deduce information from the measurements made by the detectors. Many of these techniques use structural analysis to deduce the relative spatial positions of the atoms and/or their respective movements. In other cases, the chemical nature of the elements present in the sample is identified. To have a relatively reliable picture of a molecule, it is always necessary to use several different and complementary investigational experimental techniques. Sometimes the sensitivity of the measurement is far superior to the theoretical possibilities of interpretation: very precise, reproducible data are obtained, but their relationship to the atomic arrangements defined within the molecule cannot be very well established.

The mechanics of vibrations of the water molecule are apparently easy to describe with equations, because water is one of the simplest molecules, made up of two rather different elements, each having specific physical properties. Moreover, many analytic methods are carried out using water as a solvent, because dissolved substances are examined. It is thus rather natural that the techniques invented by instrumentation designers in the course of time should in large part have focused on water for the purpose of describing the spatial arrangement of atoms. We can compare the chemist's need to know this geometry to that of the city planner who must know how the streets and squares in a city are arranged. Thus chemists compose a mental map for their personal use which they need for reasoning things out, for the creation of models, and give free reign to their imagination. Water is one of the most elementary building blocks of this construction kit, but it is necessary to understand that real components cannot usually be perceived physiologically. They appear only as ideas or images.

As a matter of fact, almost all modern instruments produce *images*. It is more difficult to translate these images onto the printed page than it is to represent formulas or tables of figures that play such a large part in the science journals traditionally used to communicate scientific information. Images often require the three dimensions of stereoscopic vision or the animation of dynamic models that only video can provide. In fact, *molecules are both rigid and in motion*, that is, the atoms of which they are composed oscillate in space about equilibrium positions and they are constantly moving by vibrating, twisting, and rotating. Because their wavelengths have the same order of magnitude as interatomic distances, x-ray diffraction allows the average positions occupied by the atoms in a rigid medium to be measured, provided that these positions correspond to the knots on a geometric meshwork, or lattice, as in the case of a crystal. If the arrangement is not regular, as is the case with liquids, glasses, and amorphous materials, it is possible by means of statistics to obtain only the average interatomic distance without any indication as to the actual spatial distribution of the atoms. Other techniques, such as measurement of wavelengths of absorbed infrared radiation, are sensitive only to atomic movements: only motion is detected by these techniques. If it is possible to use a structural model to calculate the mechanics of the molecule, the result can be compared with experimental results, and the agreement between the theoretical and experimental results can be improved step by step. Certain spectroscopic methods (primarily the absorption and emission of electromagnetic radiation) allow very precise experimental results to be obtained and establish sufficiently simple molecular structures that can be numerically treated with good precision. Naturally, as with x-ray diffraction, the possibility of comparing pre-dictions made on the basis of a structural model with the actual measurements, or their *simulation*, depends on the calculating power of computers.

HOW DOES A WATER MOLECULE LOOK?

The structure of the isolated water molecule was discovered in 1956 by a sophisticated infrared spectroscopy study designed to obtain the parameters that determine the molecule's mechanics of vibration and rotation. The use of calculations to reproduce the position in the energy spectrum and the intensity of around 2500 absorption lines in the vapor phase observed between 1.25 and 4.1 μm for each of the three molecules H_2O, HDO, and D_2O led to the determination of their shape. It is a like a compass from geometry class with its axis occupied by the oxygen atom and its two points resting on the two hydrogen atoms. The angle between the two arms of the compass is $104.25° \pm 0.05$, the length of the arms between atomic nuclei is 0.09572 ± 0.0003 nm. (The nanometer, or 10^{-9} meter, is a unit of measurement often used in crystal-lography.) The molecule is symmetric with respect to the bisector of the H–O–H angle.

Theoreticians can also evaluate the shape of a mole-cule by deducing from calculations the spatial arrange-ment of the molecule's atoms that corresponds to the lowest possible energy level. To do this, they initially consider that the atoms are far apart, with their electron clouds well-separated. Then, on paper, they observe how the electron density, that is, the probability of the presence of electrons, evolves as the nuclei get closer together until intramolecular distances are reached. These models depend on the detail of the bases chosen for the calculation, particularly the number of electron configurations considered around each nucleus. In fact, each of the electrons orbiting around a nucleus is characterized by four exclusive quantum numbers (the Pauli exclusion principle). For a nucleus-electron group that contains several electrons, the set of specific values for these quantum numbers defines an electron configuration. But for the same total number of electrons there are several possible electron configurations, because the quantum numbers associated with electrons *in situ* can change. In fact, the number of electron configurations is limited for practical reasons. There is one configuration with the lowest energy, which is thus the most stable, but each molecule (or each atom) is associated with a sort of ladder having rungs that are not equally spaced, with each rung corresponding to a possible electronic configuration.

The lowest rung, on ground level, is the basic configuration, and the other rungs represent excited configurations. If a molecule could be experimentally placed in an excited state by giving it sufficient energy, its chemical composition would not change, but its reactivity (its chemical properties) might vary considerably. In a complex (polyelectron) system, such as a molecule, the lowest energy state, the *ground state*, can correspond to combinations of a particular configuration with several *excited configurations*. In general, the greater the number of configurations (using modern methods, there are *several thousand*), the more closely the simulation calculations approach the experimental measurements. However, this process is limited by the power of computers and the cost of computer time.

For molecular calculations, water with its ten electrons is a relatively simple case, and the theoreticians may consider that each of these calculations is capable *a priori* of precisely simulating almost any of the properties of the isolated water molecule. This, in fact, is true for the distances and the angles as well as for the mechanical properties associated with the vibrations and rotations of the molecule. The water molecule can be regarded as being composed of three balls: one heavy (the oxygen nucleus) and two light (the hydrogen nuclei), linked by flexible springs, resulting in a vibrating mechanical system. The art of the calculation consists in analyzing the movement by using the classical model of the harmonic oscillator. The purpose is, on the one hand, to determine the energies associated with the modes of fundamental vibration (known as normal modes), thus to determine their frequencies; and on the other hand, to evaluate the force constants associated with the "springs" that represent the chemical bond. The operator, the mathematical object that allows the energy of the system to be found, is known as the *Hamiltonian*.

On the scale of atomic dimensions, the solutions provided by the application of the operator to the system (by using the Schrödinger wave equation for the electron energy levels) are *quantified*, that is, the solutions are discrete and can only indistinctly cover the possible energy values. This is true for the entire energy spectrum, from the low energies (relative to the ground state at rest) associated with rotations and with bond stretching, to the much greater differences in energy that exist between the ground state and excited state electron configurations. Typically, in the case of water, the energies involved for rotational and bond stretching are in the radio or infrared portion of the electromagnetic spectrum, while for the excited states these energies go up to the far ultraviolet. For low energies (I.R.), the excited states are those allowed by the excitation of the *mechanical states* of the molecule in its ground state, but for the higher energies, it is the electronic state itself that changes.

The Hamiltonian used for determining the energy differences between electronic states is applied to *wave functions* that describe the behavior of each of the electrons in the space surrounding the nucleus, with all the possible choices of quantum numbers, and the square of the wave function corresponds to the probability that these electrons will be present. These wave functions can be classified according to their symmetry properties. For example, in the case of water, there is a central axis of symmetry. The geometric elements of the molecule (axes of rotation, planes of symmetry, inversion center) determine the number of possible fundamental vibrations (normal modes) for a system of bound atoms. The reasoning used by the chemist to determine not only the number of modes but also the geometry of the atomic shifts corresponding to each mode depends on the application of *group theory* to problems involving molecules. Not only the number of lines that can be observed in the experimental spectra but also the intensity of these lines depend directly on the symmetry of the molecule and, as a result, make it possible to determine the symmetry of the molecule by simple measurements. Group theory was first used in chemistry in the 1930s. This advance profoundly changed the language of chemistry by introducing mathematical expressions that allow a problem to be condensed; they also simplify the calculations through the use of deductions made from a great deal of experimental data provided by spectroscopy.

The elements of the groups are operations of symmetry that leave the molecule unchanged. In all cases, the objects to which group theory is applied are *actions*. The simplest model is that of permutations of the digits 1, 2, and 3: the six elements of the group are 123, 321, 213, 132, 312, and 231. To make the connection between the permutations and the symmetry groups, it suffices to attach a label with a numeral to each of the atoms of the molecule. The rules imposed by the application of group theory determine the probability of transitions from one molecular state to another and thus determine the intensity of the absorption or emission that can be observed during an interaction with electro-magnetic radiation.

EXCITED STATES OF THE WATER MOLECULE

Modern chemistry uses calculations with the assistance of the group theory that sprang so romantically from the pen of Évariste Galois (1811–1832) just before his death. As pointed out above, this theory has had an enormous influence on the vocabulary of the chemist, because it allows experimental objects (such as spectral absorption bands) to be named and identified that cannot be distinguished at first glance. More or less associated with dynamics of movement and change, these "labels" (a combination of numerals and Greek letters) are part of the description of the molecule and, in particular, allow its "attitudes" to be identified. These attitudes are at the heart of the important question of "*excited states.*"

We can better understand this if we imagine that the molecule (or neutral atom or ion) lives in a building. If left alone, it stays quietly on the ground floor in its ground state, but if there is an external perturbation that supplies energy, the molecule rushes to the elevator to climb directly to the labeled floors, then goes to the windows, from which it may jump down, with the emission of a photon (luminescence). But most of the time, the molecule quickly descends the stairs, losing the acquired energy in the form of heat. This exercise is a perpetual one, because there is always something that will perturb the molecule, such as heat, light, or radiation.

A distinction is made according to degree as represented by the lower floors, the highest floors, and even the tiny landings. The interaction of water with its environment leads to constant shifts, from one landing to another, in the "building." For a molecule spread all over the world, from the ground to the stratosphere, the consequences of these "jumps between floors" are immense for the mechanisms of life, such as those studied in climatology. If we come down from this cosmic level to our everyday level, we can see that even the most domestic of tasks also exploits the properties of the excited states of water. The existence of these excited states is actually as important as that of the entities that sustain them: the neutral atoms, ions, and molecules which constitute matter from one end of the Universe to the other. Without these excited states, no *communication* would be possible, for example, through the absorption or emission of light—there would be no exchanges! Without these discrete, effortless, and sudden jumps in quantum numbers, the world would be just an inert pile of protons, neutrons, and electrons.

There are *three* normal modes of vibration for the water molecule. *Symmetric* stretching, during which the oxygen and hydrogen atoms simultaneously move closer together and further apart, is found in the infrared spectrum at 3656.65 cm^{-1} (the cm^{-1} is a unit used for energy measurements). This figure quantitatively expresses the height in the "building" of the "floor" to which the excited state corresponds; its wavelength is 2.73 μm. *Antisymmetric* stretching, during which the two hydrogens have opposing movements, is found at 3755.79 cm^{-1} (2.66 μm). And, finally, a *twisting* movement, which results in absorption, occurs at 1594.59 cm^{-1} (6.27 μm). In the case of heavy water, because the masses involved are different, although the geometry and the distances are nearly the same, the energies associated with vibrations show a marked change (respectively 2671.41 cm^{-1}, 2788.05 cm^{-1}, and 1178.33 cm^{-1}). Excited states are temporary states: the molecule retains the energy borrowed from the electromagnetic wave for only a certain time, then the energy dissipates little by little in the form of heat and the molecule thus returns to the ground state. The number of molecules that "de-energize" per unit of time is constant. The rate of this process is expressed by a statistical measurement that represents the time necessary for 63% of the molecules to return to the ground state. For the symmetric vibration state of water, the half-life is 240 milliseconds (ms), and is 13 ms for antisymmetric vibration.

The rotational modes are superposed on the vibrational modes by coupling. The differences in energy between the vibrational modes and the rotational modes are slight, and are located in the range of waves having centimeter wavelengths. Determination of these energies makes it possible to calculate the moments of inertia of the molecule. This property is used in the kitchen to heat water in a microwave oven. Electromagnetic energy is absorbed to maintain the rotations of bonds, and the excitation energy thus provided is dissipated by the system of molecules in the form of heat. Absorption at 2.45 gigaherz (GHz) is specific to water. The infrared absorption spectrum of water is rather high on the energy scale as compared to many other molecules. Since water vapor is an important component of the Earth's atmosphere, its presence limits the ability of infrared radiation to penetrate to the ground through the narrow "atmospheric window" located between 8 and 11 μm, where water does not absorb. To study the emissions of stars outside this window, it is necessary to make astronomical observations from orbiting satellites.

Water vapor also absorbs energy in the far ultraviolet at a wavelength of 0.165 nm. For water condensed in the liquid form, this absorption is located higher in the energy scale, at 0.147 nm. For the isolated molecule, there is one absorption band whose energy is 7.49 electron volts (1 eV = 8.066 cm^{-1}). Other bands (at 9.75, 9.81, 10.00, and 10.17 eV) correspond to the position of excited configurations for which an electron has changed quantum numbers. Here we touch upon another property of water whose practical, economic, and social implications could one day become important. Water can be dissociated by the action of light. If water vapor is subjected to ultraviolet radiation, it rapidly dissociates into H and OH, starting from its excited states and their vibrational levels. It is hoped that this method can be used to obtain a nonpolluting fuel—hydrogen. No pollution is produced when hydrogen burns because the only by-product is water! Thus, we would have a truly renewable "natural" energy source if we could succeed in "cracking" water, for example, by using solar energy. The hope of one day building an economy based on hydrogen has stimulated considerable research in the past with regard to methods capable of decomposing water. Although many ingenious cycles have been discovered, these techniques seem to have been hindered by their lack of cost effectiveness, as long as petroleum cracking remains less expensive.

The large gap between excitations in the infrared and excitations in the far ultraviolet explain why water is *transparent* to visible light. Ideally pure water is, in theory, colorless. Deep water appears blue because it absorbs the red wavelengths due to the harmonics of its infrared bands. In general, the color of an object is related to the absorption of sunlight by the excited state of a solid, liquid, or gas. A surface that absorbs all of this light appears black, while a surface that reflects all light appears white. A surface that absorbs only part of the visible spectrum, as is most often the case, reflects the complementary part of the spectrum, and has color. For example, because the chlorophyll in leaves is green, it has an excited state in the red portion of the spectrum. These rules are valid not only for visible light but also for other parts of the electromagnetic spectrum.

THE ELECTRONIC STRUCTURE OF THE MOLECULE

When theoreticians study a molecule (or an atom) containing several electrons, in order to perform the basic quantum calculation they divide the electrons into what are known as delocalized *orbitals*. These orbitals are classical mathematical functions (spherical harmonics which may have imaginary coefficients) describing the position (distances and angles) of electrons in space which have the symmetry of the molecule. To obtain a visual representation of this electron world, the chemist prefers to manipulate these crude orbitals to cause the imaginary coefficients they contain to disappear, and to transform them to obtain information about the localized electron density in real geometric space. Such transformations do not affect the energy of the system and allow many conventional intuitive concepts to be rediscovered.

The representation obtained for the molecules is a sort of three-dimensional map with curves for the levels that quantitatively express the local electron density, which is the probability that the electrons are present at a particular point. After these transformations, the localized orbitals, whose overlapping is involved in chemical bonding, are thus essentially centered on the participating atoms, each providing an electron to the common system. For water, the localized description also shows two orbitals that have electrons which come exclusively from the oxygen, with the lobes projecting above and below the plane formed by the three atoms. These orbitals contain *lone pairs* of electrons, which are not shared with other nuclei but can nevertheless be used for bonding. For the isolated water molecule, this description points to the image of a quasi-tetrahedral shape for the electron cloud, which accords well with the classic notion of the directionality of bonds largely used to try to explain the structure of liquid water. However, it is not certain that the sophisticated modern calculations of charge density actually demonstrate the existence of lone pairs on the "back" of oxygen. This question remains unanswered and controversial: while some researchers believe it is true, others do not.

For a long time, difficulties were encountered in reproducing measurable electric properties of the water molecule when using the various techniques for calculating molecular orbitals that were fairly successful in simulating the spatial or mechanical data.

In fact, it can be deduced from the localized model described above that the distribution of electric charges in the molecule is dissymmetric: The negative charge is concentrated near the oxygen, and the positive charges are concentrated near the two hydrogens. Since the electric charges are separated, this molecule is said to have a *dipole moment*. This dipole moment can be measured by using high-resolution spectroscopy. While the rotational absorption lines are theoretically degenerate, according to the rules of group theory (two different electron states must correspond to the same energy), they are actually separated by the internal electric field of the molecule. This is known as the Stark effect, named for the German physicist Johannes Stark (1874–1957).

The existence of the effect of the electric field in the space around a dipole requires a difficult and complex mathematical treatment. Measurement of the energy difference mentioned above yields a numerical interpretation for the electric dipole moment, which is the coefficient of the first term of the Taylor series expansion for the electric field created by the dipole. This coefficient corresponds to a vector (which is actually a first degree tensor), and for two equal and opposite charges separated by a small distance it is equal to the product of the charge times the distance. The coefficient of the second term in the series (a second degree tensor with five components) is the quadripolar molecular moment, which is much more difficult to measure. For water vapor, the experimental value of the dipole moment is 1.8546 ± 0.0004 debye. Compared to other molecules, this is an extremely high dipole moment. The different wave functions yield more or less the same result, with the difference being 5 to 20%. The possibility of the existence of lone pairs has complicated the model for electric charge distribution in the water molecule. In 1951, the Danish scientist N. Bjerrum constructed a model in which the charges were concentrated at the tips of a tetrahedron and consisted of two negative charges on the oxygen side and two positive charges on each of the hydrogen sides.

V

THE COLLECTIVE LIFE

OF WATER MOLECULES

With its three centers of mass and, perhaps, its four centers of electric charge, the isolated water molecule is a rather complicated chemical entity. It is easy to see that this complexity can reach limits that are difficult to determine when the isolated water molecule is plunged into a large collection of other water molecules which, for purposes of abstraction, we have until now considered to be isolated. This collectivization occurs when water vapor condenses to liquid or to solid water. When one water molecule approaches another, their respective electric fields affect each other, and changes occur. The ground-state electron configurations of some metals are different in the vapor state and in the condensed state. In the case of water, the mechanical (spectral) microproperties are only slightly affected, but the complexity of the electric structure leads to a special behavior for molecules in collective states.

To understand what happens when previously separated molecules are joined, imagine that the isolated molecules now in motion are like beautiful butterflies beating their wings. When there are many butterflies in an enclosed space, their beating wings can overlap and interfere with each other if the butterflies approach each other too closely. To allow them to gather collectively, it would be advantageous for the butterflies to have a second set of wings, which are shorter or arranged differently in space, so that each butterfly could fit better into the great mass of butterflies. According to the logic of this metaphor, the "wings" of the butterfly represent the distribution of electric charge in the water molecule.

Researchers are attempting to understand the collectivity of water molecules gathered in large masses, whether they be liquids or solids, from the experimental results established in very small aggregates. This *recent* triumph of science uses feats of experimental technique to produce and examine small clusters of water molecules. Although the number of atoms involved is quite small, unexpected properties fundamentally related to the flexibility of chemical bonds have been discovered. This step is indispensable to understanding how water molecules arrange themselves within a mass of liquid. There is not the slightest doubt that such information is of the greatest interest if we realize that the chemical or physical properties associated with liquid water depend on these structures, and that this is the medium in which the chemistry of the Earth's environment, especially the chemistry of life, occurs.

DIMERS, TRIMERS, AND COMPANY

The previous chapter presented a portrait of the isolated water molecule. This is obviously almost an ideal case, since water usually leads a collective life; like congregating with like to produce giant clusters that cover the Earth as oceans. Even in the form of vapor, the water molecule cannot be entirely free of all bonds. We know how to pair water molecules by using molecular jets that allow associations of molecules to be created. By imposing a very high kinematic viscosity on a gaseous flow that contains monomers (single molecules), they are forced to constantly collide with each other. To achieve this, the gas is forced through a small opening (less than 1/10th of a mm) in a sort of cone-shaped nozzle from which it is expelled at a rate two to ten times the speed of sound. The flow is released all at once into a vacuum within the apparatus, which stabilizes the successful associations by limiting the chances for destructive collisions. One portion is selected by the diaphragms that lead into an analysis chamber where spectroscopic or even structural measurements can be made, for example, by electron diffraction, if the fragments produced are large enough. The size of fragments ranges from the association of two individual molecules (dimers) up to clusters or aggregates of several dozen molecules. Beyond the association of three individual molecules (trimers), some difficulty is encountered in comparing a structural model based on the interpretation of spectroscopic measurements (usually in the infrared) with the calculations of molecular mechanics, which soon become very complicated. For example, the trimer of water has twenty-one degrees of freedom!

This is the picture of "double water"—the dimer of water, or $(H_2O)_2$—which we know well today. The official version of the portrait corresponds to the predictions: the two H–O–H angles of the pair retain their value of 104.5°, and a bond (with a length of 0.096 nm) is seen on one of the two molecules connecting the oxygen atom to one of the hydrogens, which is aimed at the lone electron pair of the oxygen of the other molecule, forming a straight line O–H–O with a distance of 0.299 nm between the two oxygen atoms. This model demonstrates *hydrogen bonding*. However, the orientation of the three hydrogens that are not on the straight line appears to be eminently variable, because there are at least eight equivalent geometric forms for the water dimer! This question of conformation is further complicated because the potential energy calculations demonstrate that the linearity of the hydrogen bond can suffer from remarkable distortions. It is almost as stable when the line bends at an angle of 50° at the level of the hydrogen, and its length can vary a great deal—from 0.19 to 0.31 nm! Among all the potential structures, there is even one for which a single oxygen atom supports as many as four hydrogen bonds, so that the hydrogens are found halfway between the two oxygen atoms! The energy barriers calculated for these different geometries are extremely low, not exceeding 0.87 kcal/mole, so that the simple effects of thermal agitation allow the molecule to pass readily from one molecular form to another. This extraordinary flexibility presents particular obstacles to analysis and simulation by calculation when it is desired to study the properties of water molecules in the collective state. The experimentally determined value for the energy of formation of the hydrogen bond in the linear dimer is 3.59 ± 0.5 kcal/mole at 273 K (0°C), a rather high energy required to stabilize the dimer relative to the room temperature thermal fluctuations that prevent it from decomposing too rapidly into two separate water molecules.

The abundance of water molecule dimer forms drives the structural chemist to despair. In fact, each form of the dimer can be associated with a unique "excited state" geometry. Above we used the metaphor of the building with different floors to describe the role of excited states in the molecule's absorption of energy from the external medium. In this case, the energy differences are so small among the different molecular forms that to go from one form to the other it suffices to draw only on the permanent energy reservoir provided by the ambient temperature! Here we no longer have one building, but rather a collection of small houses at the same level! The structural chemist is thus now in a predicament, having the feeling that the rigid energy hierarchy of states no longer applies. If everything were on the same level, everything would be possible. It is precisely this versatility that constitutes the originality of the water molecule bound in the collectivity with all its fellow molecules. The theoretical predicament is further compounded by a formidable technical problem. We can gain some idea of this from the butterfly metaphor proposed above: the butterflies juxtapose themselves in many ways, somewhat like a jigsaw puzzle consisting of soft, moldable pieces whose solution yields different images.

The theoretical distortions predicted by the models for the water dimer are supported by experimental measurements, since it is possible to study the dimer with a variety of techniques that allow any structural fluctuations to be observed. The water dimer can be trapped at low temperature and studied while imprisoned, isolated in solid crystalline matrices formed by a network of atoms of rare gas, such as neon, which is transparent over a wide range of wavelengths and thus greatly facilitates spectroscopic studies. These studies reveal that the energy associated with the stretching of the hydrogen bond corresponds to an infrared absorption band at 240 cm^{-1}.

The neutral polymer species of water have cousins: clusters of water molecules that carry a *positive* electric charge (resulting from proton capture) or a *negative* electric charge (resulting from electron capture). These clusters are particularly easy to study by using modern ion "screening" techniques that measure the ratio of the mass of the cluster to its electric charge. To isolate and count the charged ions, techniques involving deflection, trajectory curvature, and the tracking of charged particles are used which are based on the manipulation of electric and magnetic fields. This technology forms the basis of one of the most powerful analytic techniques, namely, mass spectrometry, which is used to determine the "weight" of the molecules; its great sensitivity permits differentiation of charged particles formed from different isotopes. Thus, a beam of large water ion clusters is selected to be directed toward a cavity where it can be specifically examined by spectroscopic techniques. Using a continuous infrared laser beam, the wavelengths of the radiation that will break up the cluster can be precisely determined. The mass spectrometer identifies specific molecular fragments, signatures generated when the explosion (dissociation) of the complex ion is caused by vibrational excitation. The range of mechanically excited states of the cluster of bound water molecules is thus established. A high degree of precision in measurement can be achieved for an extremely narrow range of wavelengths. For hydrated protons, this technique is used in the region corresponding to the stretching of the O–H bond located between 3550 and 3800 cm^{-1}. These hydrated protons are produced by means of electric discharges emitted in an atmosphere under pressure containing a low concentration of monomer molecules; the water molecules absorbed on the surface of the walls of the metal chamber will suffice. The plasma is formed by injection through a nozzle a few dozen microns in diameter, and is released at high velocity into the vacuum of the measuring device.

These are not sophisticated operations with only a purely theoretical interest. On the contrary, they are important for understanding an ordinary phenomenon that is of obvious practical interest: transporting an electric current through water. The way the charges circulate in the liquid medium and especially what physical vector or material carries the electricity remain obscure. Research on this subject lies in the areas of *electrochemistry*, a discipline that is developing at the crossroads of industry and biology. One of these embodies a hot topic in current research, namely, the interactions between the liquid (electrolyte) and the solid interfaces that capture the positive or negative charges (electrodes); these relationships must be understood before valid models for the reactions can be established. This research only allows hypotheses to be formulated: electrochemists have identified several potential candidates, several ions that could participate in the flow of electricity—particularly in the transport of protons—in a liquid medium.

Aqueous charged species can also be found outside the laboratory. Clusters of water molecules can capture a proton, H^+, resulting in the formation of $H_3O^+ (H_2O)_n$, known as *hydrated protons*, which predominate in the D region of the Earth's mesosphere, the layer located at an altitude of between about 30 and 55 mi [50 and 90 km] above the surface of the earth. This is a recent discovery (1965), because in order to understand the role of these unexpected molecules it was necessary to wait for the results obtained from the use of mass spectrometers on board space probes and satellites. The density of particles in this layer is low (on the average, only 10^3 per cm^3). This explains the stability at this particular altitude of these fragile species, which are relatively protected from the photochemical effects of solar radiation; they are destroyed only by chemical reactions occurring in low-density media, rare collisions with another molecule or atom (such as ammonia or nitrogen oxides), or encounters with electrons or negative ions. The positive ions corresponding to $n = 0, 1, 2, 3$, namely (H_3O^+), $(H_5O_2^+)$ $(H_7O_3^+)$, and $(H_9O_4^+)$ are the most abundant. In the cold layer (180 K = −93°C) known as the mesopause, located at about 53 mi [85 km] above the Earth, n can reach values as high as 10.

Hydrated protons are not just curiosities that exist only in the atmosphere. They are also found in some crystalline structures, such as hydrated organic acids or other materials that contain water in their chemical formula. Their study makes it possible to determine the form of these associations, and it is found, surprisingly, that they have different geometries. Theoretical calculations that compare the possible forms and determine their energies provide a guide for interpreting these different forms. These calculations confirm that the same formulation can correspond to several different potential symmetries, which can be visualized as a proton surrounded by water molecules (for example, $H_2O–H–OH_2$) or as an association of the hydronium ion H_3O^+ with one or more water molecules. Actually, both of these possibilities can occur at the same time: ultimately, the proton can easily oscillate from a narrower association with one of the water molecules to form a sort of pyramid with ternary symmetry (H–O–H angles close to

120°), until it plays the role of a nearly flat bridge mid-way between two water molecules.

The water molecule has a great affinity for the proton (167 kcal/mole) and the resulting hydronium ion has the same number of electrons as ammonia, NH_3. Like ammonia, its umbrella-shaped pyramidal structure is susceptible to being turned upside down and reversing in situations where the energies are situated in the far infra-red (55 cm^{-1}). This hydronium ion also readily combines with another water molecule, resulting in the species for which $n = 1$ ($H_5O_2^+$) and the bonding energy, which decreases as n increases, is 31.6 kcal/mole. The energy differences between the potential geometric forms corresponding to the same chemical composition are only on the order of 0.2 kcal/mole! This energy is very low (70 cm^{-1}), below the average thermal energy at room temperature (208 cm^{-1}). Consequently, *as is the case with the water molecule dimer, the hydrated protons pass freely from one geometric form to another.* Thus we discover an energy factor that facilitates deformation and movement; this does not greatly concern the structuralists, because they prefer to manipulate well-identified species rather than molecules that resemble soft marshmallows!

The negatively charged clusters $(H_2O)_n^-$ have almost the same properties. The water molecule dimer weakly binds an electron, the binding energy being estimated at about 20 meV (or 160 cm^{-1}). On the other hand, larger clusters ($n > 10$) can hold the electron more strongly, with binding energies that in this instance are much higher in the electron volt (8066 cm^{-1}) range. For sizes between $n = 11$ and $n = 32$, the electron is retained in a surface state, whereas between $n = 32$ and $n = 64$, the electron remains buried inside. The geometry of these different species appears to be as open and variable as that of their positively charged companions.

The study of these little clusters of molecules marks an important stage in our understanding of what occurs in condensed matter, a medium that is much more difficult to study because it brings together infinitely more actors than we have previously considered: there can be as many as 10^{22} atoms per cm^3 in a solid or in a liquid! Many of the characteristics of potential chemical bonds between water molecules, as demonstrated by the study of small aggregates, are found in the most familiar forms of water, which intensifies the interest of researchers in the exotic species described above.

ICE

With its small clusters, ice has long been the best-known form of the collective gathering of water molecules. This is so because ice can be analyzed with the formidable arsenal of experimental characterization techniques available for studying the chemistry of solids. Ice has historically provided the basic model that has guided research on the local structure of liquid water, a model visually confirmed by the macroscopic appearance of the melting ice-packs at the poles during the summer. This model has always been attractive to researchers because it is intellectually reassuring to be able to depend on a well-known arrangement, with a beautiful geometry, that favors not only logical reasoning but also simplified calculation. Unfortunately, this conventional model is perhaps no more than a projection onto nature of our love of order.

Ice forms crystals. In a crystal, the atoms are in an ordered arrangement: they occupy positions that repeat regularly at the intersections of a geometric network, or lattice. The smallest repeating unit in three dimensions is the unit cell. The elements of symmetry of this unit cell (planes, axes, points) allow a crystal to be classified in one of 230 possible symmetry groups. In contrast to crystals, other solid materials, such as glasses and amorphous materials, are composed of atoms whose positions do not regularly repeat on a large scale in space; thus a unit cell cannot be defined. Snowflakes demonstrate that the most ordinary ice belongs to a group with hexagonal symmetry. This is only one of the structures of ice, known as Ih. There are at least a dozen other structures that can be obtained by varying the temperature of a sample from 0 to –200°C and its pressure from 0 to 25 kb. *As the pressure is increased, the melting point of ordinary ice decreases*. This explains why the ice skater glides so easily on the ice: the pressure exerted by the blade causes a small amount of ice to melt, which releases a thin film of water that lubricates the blade. If a thin wire with a weight on each end is placed on top of a block of ice, the ice never breaks while the wire is slowly passing through it, because the water formed under its pressure refreezes after the wire has passed through.

To enter the world of crystallography, we need only to turn on our visual imaginations. Crystallographers let their thoughts roam through the architecture of absolute and perfect spaces making up fundamental polyhedra. In this case, the imagination of the reader should focus on the tetrahedron, which, by definition, has four vertices. Its four faces are equilateral triangles, and there are a total of six edges. From the center of a polyhedron, an edge is "seen" at an angle of 109°28'. Imagine that one atom is placed at the center, other atoms are at each of the vertices, and an infinite number of tetrahedra are piled on each other, vertex to vertex.

The structure of a crystal is determined by numerical analysis of x-ray and/or neutron diffraction data. In the case of ice, the arrangement of the oxygen atoms is easiest to determine because they form a perfect tetrahedral arrangement. Each oxygen atom is 0.2751 nm away from its four oxygen neighbors (at 77 K) and 0.45 nm away from its twelve neighbors in the next row. Such a geometry suggests that ice is a three-dimensional network of water molecules connected to each other by hydrogen bonds. Since the ideal tetrahedral angle is 109°28', we might wonder whether the condensed water molecule has the same shape as the free molecule whose bond angle is 104°5'. X-rays are diffracted by external electron layers, so that hydrogen is almost invisible, whereas the neutrons are diffracted by the nuclei of larger atoms. Neutron diffraction of D_2O shows that the O–D distance in Ih ice (0.101 nm) is slightly greater than it is in the isolated molecule (0.09584 nm), and that the D–O–D bond angle is perhaps only one degree greater than the 104°5' in the free molecule. The hydrogen bond is thus not perfectly linear in ice: hydrogen is not located exactly on the line that connects the two oxygens. It is estimated that the distortion is on the order of 2°.

The structure of ice is thus dominated by the arrangement of the oxygen atoms. Each oxygen atom has four symmetrically equivalent equidistant oxygen neighbors. The oxygen is said to have a *coordination number* of 4. Each O–O segment, 0.275 nm long, is associated with a hydrogen atom. But the covalent bond between each hydrogen atom and one of the oxygens, which is characteristic of the water molecule, is short (0.1 nm). Hydrogen is thus not found in the middle of the O–O segment: it is found either to the right or to the left! Since the angle of opening of the water molecule is close to the tetrahedral angle, it is easy to demonstrate, by topological reasoning, that in relation to *one* oxygen atom there are *four* different ways of arranging the hydrogens in *each* water molecule, thus insuring hydrogen bonding with each of the four neighboring oxygens. As a result, the *network of hydrogen atoms in the ice structure is disordered*. This would not be the case if the hydrogen could be placed precisely in the middle of the O–O segment. The corollary to this structural disorder is that the electric dipoles of the molecules are also disordered. The average dipole moment per molecule in ice (2.60 debye) is much higher than it is in liquid water (1.84 debye).

The same basic topological structure accounts for the other types of ice that occur at high pressure as mentioned above. But the tetrahedra are more or less deformed, and the distances vary. This, in particular, is the case with the hydrogen bond, which stretches. Ice Ic, one of these types prepared by condensing water vapor on a cold surface maintained at a temperature below –80°C, forms cubic crystals with the diamond structure having many elements of symmetry. Under the same conditions, an amorphous type of ice (in which the oxygens no longer form a regular network) can be obtained by slow deposition at a rate of 1 μm/hour at temperatures below 135 K (–138°C).

Neutron and x-ray diffraction, which are generally well-suited for determining structures in condensed matter, are not effective in difficult cases, such as that of water. In the case of water, chemists must also resort to indirect methods, that is, experimental results whose theoretical simulation shows that they depend directly on the arrangement of the atoms in space. It is in this way that spectroscopy data increasingly provide the basis for choice arguments for the construction of models. However, they also provide grounds for potential disputes, ammunition in the structural battles often fought by researchers over the different models.

An absorption or emission spectrum is composed of *sharp lines* if the molecules or ions occupy *well-defined* crystallographic sites and if they are far enough away from each other (for example, separated by dilution with inert chemical elements). The geometric elements of symmetry at the site and the rules of group theory indicate the number of lines that can be observed. Conversely, observation defines the symmetry of the site, and, in a solid, different types of spectroscopy (such as radio, infrared, visible, or ultraviolet) frequently serve as "punctual probes" to obtain local structural information from optical spectra that are dependent on the symmetry of groups of molecules or ions.

Stretching of the O–H bond, located between 3000 and 4000 cm^{-1}, appears as sharp lines in solids containing OH^- ions linked to cations that are located relatively far apart, provided that the geometry of the O–H bond occurs regularly throughout the crystal. We can easily predict from the description of the distribution of the O–H bonds in the water molecule within the structure of ice that the infrared spectrum will not be simple! In fact, because there are many different geometric situations around the O–H group, the frequency of stretching of each one of them, which is very sensitive to the environment, can vary a great deal.

Moreover, the spreading of the lines indicates that the water molecules are not isolated—they are all touching, and so affect each other. Their mechanical movements depend not only on the internal couplings of stretching, twisting, and rotating, but also express the result of the oscillations affecting their neighbors, somewhat like the agitation of an individual in a crowd whose movements are restricted by close contact with the others in the crowd. *Inter*molecular movements, of lower energy, are superimposed on mechanical *intra*molecular movements. This explains why the infrared spectrum of ice, instead of containing lines, is composed of *relatively wide continuous bands* from which only semi-quantitative information can be extracted—provided samples are used in which the masses are manipulated by isotopic substitution (since the spectra are shifted, the O–H vibrations of H_2O dissolved in D_2O can be observed) or the temperatures are changed to limit movements and thus to shrink the width of the spectrum.

The stretching of the O–H bond can be observed in the infrared absorption spectrum of ice in the curve appearing between 3000 and 3500 cm^{-1} with two broad peaks at 3130 and 3220 cm^{-1}, followed by a shoulder at 3370 cm^{-1}. It is very difficult to use calculations to simulate a spectrum resulting from so many interactions and whose width expresses the elusive statistical distribution of hydrogen atoms in the crystalline network. The vibrations of the hydrogen bond itself form a large band in the 150-300 cm^{-1} region. In the region of the spectrum below 1000 cm^{-1} (thus above 10 μm), it is possible to obtain information on intermolecular movements (movements of molecules with respect to each other, particularly stretching vibrations and rotations about the hydrogen bond axes).

LIQUID

Although ice is the state of water that has been the most completely described, the liquid form is of course the one with which physical chemistry and biology are most generally concerned. It is the most important yet the most difficult form to understand because of formidable barriers to our knowledge. The problem is unyielding and, like a hydra, it transforms and regenerates itself, even as researchers use increasingly sophisticated techniques. The current state of our knowledge of the structure of liquid water can be appreciated by contemplating the Tofuku Ji Zen garden in Kyoto, Japan. On the left are flagstones arranged in a checkerboard pattern on a carpet of moss. Little by little, the pattern of the flagstones becomes less like a checkerboard, they lose their geometric relations and scatter, until they finally disappear to the right in the expanse of moss and roots. This suggests the "annihilation of all order in the unknowable" (according to one guidebook for Japan). Such is our knowledge of water. Some look at the left side and see the flagstones emerging distinctly from the moss, representing the arrangement of little icebergs, pieces of solid in a liquid. Others, viewing the right side, see only the disordered microstructures whose indistinct detail reveals only a fantasy of chaotic encounters. Experimental arguments and calculations can, of course, be marshalled to support either view. This symbolic story shows that nothing is ever certain, and that a complete description remains difficult, if not unattainable; we must be satisfied with different points of view. Fortunately, the Zen garden is too large to really be seen at a single glance.

At 0°C, ice melts, its crystalline structure collapses, and water molecules roll freely over each other. Here is a simple image that can correctly represent the fusion of a neon crystal, for example, because at low temperatures, atoms of noble gases act like rigid spheres. However, this image is not appropriate for water. Since models for ideal liquids, as they are usually envisioned by physicists, cannot be used to describe water, we will first examine some experimental results.

X-rays can be diffracted by a liquid, to be sure, but it is not possible to determine from this information the relative positions of the molecules with respect to each other. The only thing that can be determined is the average distance between atoms whose electron clouds diffract x-rays. (We might recall that this is not the case with hydrogen.) Using this technique, the oxygen-oxygen distance ranges from 0.284 nm at 4°C to 0.294 nm at 200°C (under pressure). The second and third neighbors appear at 0.45 nm and 0.7 nm respectively. Beyond this distance, nothing is seen, which implies that the geographic zone where the molecules are relatively ordered with respect to each other during the interaction of x-rays with matter (approximately 10^{-15} second) is limited to a radius of 0.8 nm near the melting point; it decreases to 0.6 nm at 200°C. Actually, the experimental curves on which these determinations are based reveal nothing other than the dominant statistics around the cited values. Mathematical manipulation of the data can be used to determine the average number of neighbors for each water molecule. There is relatively good agreement among the different techniques, which yield a coordination number of 4.4. The reader will recall that in ice the coordination number for oxygen is 4.

Since the number of neighbors is increased, we must assume that the packing is greater in the liquid than it is in the solid. From this structural conjecture emerges one of the great fundamental properties of water that has enormous consequences for our environment: *the density of liquid water is greater than that of solid water.* Everyday observations confirm this fact: at temperatures just above the melting point, ice floats. When water freezes, it expands and can cause the container which holds it to shatter. Scholars determined as early as 1667 that the density of water is maximal at 4°C, before thermal expansion causes the density to decrease again as the temperature increases. This detail is fundamental to life on Earth, because this property explains why rivers and lakes freeze from the surface down. If ice had another structural configuration, the oceans at the poles might have frozen all the way down to the bottom. This variation in density is a striking example of the unusual evolution of the physical properties of water that occurs with variations in temperature.

Raman spectroscopy, named for the Indian physicist C. V. Raman (1888–1970), is extensively used to study the structure of liquid water. Monochromatic radiation interferes with matter and exchanges with it energies that correspond to the range of vibrational states. Upon spectrographic analysis, lines of very low intensity are observed to the side of the direct beam, which is always intense. These weak lines have wavelengths that are higher (the most visible) or lower than that of the direct beam, and they are shifted by amounts corresponding exactly to certain mechanical vibration frequencies of the sample. In the past, mercury lines were used, but now visible-light laser radiation has proven much simpler to use. The interaction between radiation and matter that occurs during a Raman spectroscopy experiment also depends on the rules of group theory that govern not only the observable vibrational symmetries, but also their intensities. Polarization of coherent electromagnetic waves greatly facilitates interpretation of the experimental measurements.

When this technique is applied to water, two regions can be distinguished. The first consists of the *intra*molecular vibrations of water, which essentially consist of stretching of the O–H bond, and the second consists of its *inter*molecular vibrations, which represent the shifts of one water molecule relative to another, specifically, stretching of the hydrogen bond and the rotations about its axis. These studies make extensive use of H_2O/D_2O mixtures (involving the formation of a combined HDO molecule) for which the spectra, in the case of intramolecular elongation, appear in two peaks, one centered around 3500 cm^{-1} (O–H), and the other around 2500 cm^{-1} (O–D). Both of these can be resolved into four components by Gaussian decomposition of the experimental curve (the four maxima occur at 3225, 3415, 3565, and 3640 cm^{-1} for H_2O). The results are interpreted as indicating the simultaneous presence of O–H bonds that support a hydrogen bond and those that do not. Raman spectroscopy, especially in the region of intermolecular vibrations below 1000 cm^{-1}, yields a model for the spatial arrangement of the water molecules when they are linked together in the liquid. It has been possible to draw conclusions from the characteristics of the spectra with regard to the symmetry of the molecular entities that produce them. The results suggest that *a group of five water molecules connected by hydrogen bonds is formed*, one in the center, with the other four oxygens arranged in a tetrahedron around it. The hydrogens are distributed in such a way that the group of five molecules exhibits a total symmetry whose elements are one axis and two planes containing the axis that are perpendicular to each other. Clearly, this is just like the structure that occurs in ice, constituting the basic unit in which the water molecules are present in the liquid.

By assuming the existence of this entity, it is easy to imagine that the evolution of the properties of water that occurs with changes in temperature corresponds to the progressive decomposition of this primary bond found in ice. Near the four hydrogen bonds, on the average, there will be three, then two, then only one water molecule, and finally all the water molecules will be released from their participation in hydrogen bonding as the ice melts. The molecules are no longer constrained to orient themselves with respect to each other. With this model, it suffices to determine the amount of each species, to select one value for the energy associated with breaking the hydrogen bond (2.5 kcal/mole), thereby reproducing both the absolute value and thermal evolution from melting to boiling for measurable macroscopic quantities. The experimental value for specific heat at constant pressure is 18.2 cal/deg mol at 0°C and the dielectric constant is 87.92 at 0°C, which are both relatively high values as compared to those for other liquids. Naturally, with a sufficiently large number of parameters that are freely adjustable, and are themselves inaccessible to experiment, there is quite a risk that many measurements could be arbitrarily simulated. The model (0, 1, 2, 3, 4) for hydrogen bonding can be used as the typical example of a "mixed model" for water.

Other models have been proposed by researchers who consider the relatively open structure of ice to be capable of hiding "free" water molecules in the interstices of the network during its collapse at the melting point. Such a hypothesis also accounts for the information provided by Raman spectroscopy; moreover, it offers a simple explanation for the initial increase in density that occurs with the increase in temperature since, at the beginning, the "holes" of the ice network that is in the process of breaking up become filled. There is a third school of thought which holds that it is sufficient to strongly twist the bonds in the ice—which, after all, could not depend too much on the orientation, as is the case with the water dimer—to obtain a uniform statistical mechanical representation for the structure of the liquid, without the necessity of imagining that it fragments into different entities. This model has the advantage of accounting fairly well for the measurements of the distances between neighboring oxygen atoms as determined by x-ray spectroscopy.

It will be remembered that 4.4 is the average number of neighbors for an oxygen in liquid water as determined by x-ray spectra. Under these conditions, some water molecules have *more* than four neighbors! A new theory, proposed in November 1991, hypothesizes that the exceptional mobility of water (that is, the ease with which it flows, especially under pressure) can be explained by a model having a mixture of fourfold-coordinated and fivefold-coordinated species. For fivefold-coordinated species, *one* lone electron pair of the oxygen atom accepts *two* protons from two other molecules. Such a situation would change not only the local density, but also the mobility, because groups of five molecules would move faster than groups containing only four molecules. This mechanism could explain the incessant reorganization of the hydrogen bond network. This network cannot remain rigid, although the average thermal energy available is less than the thermal energy necessary for destroying or forming one of these bonds.

The different conventional models for liquid water, whether they are "uniformist," "interstitial," or "a mixture," all share the same belief in a *structure* of liquid water. The example of ice is so pervasive, so omnipresent, and so active in the imaginations of researchers that it is difficult to escape from the idea that there is an edifice somewhere—one or more basic constructions, a topology—under which everything must be subsumed and from which all the properties can be derived by incremental modifications. Wanting the structure to be this way is equivalent to attempting to "think" the motion of the molecules into stopping—to freeze the random rapid movements of the molecules relative to each other in order to use the components to create a local structure. The image could then be stopped, examined, understood, calculated; and repetition would provide the key to the properties of the whole. The elusive dynamics of exchanges and contacts—disturbing to the modern structuralist mind—as well as the flexibility of the curves and folds are neglected in the frenzied research for geometrically analyzable atomic arrangements. However, today we know how to use animation to create other images of liquid water.

SIMULATION

There are no references to *experiments* nor any subtle instrument-generated data to interpret what we are about to describe. We will attempt to understand the "structure" of liquid water by entrusting the entire problem to the good offices of a computer and determining whether the computer-generated solution allows us to reproduce and simulate the numerical values measured for the *macroscopic* properties of water, especially their evolution with temperature. The computer will transform the mathematical formulas into images and finally show us the molecule as it "lives" within the liquid, especially how its situation evolves over time in relation to its neighbors. To obtain these images, we must first construct a powerful calculation tool.

The *ab initio* calculations (those that rely only on mathematical considerations) for the structure of water yield a good description of the monomer, describe the dimer reasonably well, but become astronomical when it comes to the trimer! In this case, even if the most stable form having the lowest energy could be determined, it would be necessary to consider those that have a slightly higher energy level in order to simulate the properties as a function of the variation in temperature. The theoretical problem is thus quite difficult. In principle, within the framework of classical statistical mechanics, we can estimate the sum of different potentials that cause the particles to be attracted to or repulsed by each other, and examine the bonds between dimers, trimers, etc. This is known as the *n*-body problem, but for the calculation to have any meaning, the value of *n* must be at least several hundred. In the case of water, it is difficult to consider all the potentials involved, such as the directionality of the hydrogen bond, the multipolar electric forces, and so on, the more so since it is necessary to integrate the equations over the number of *degrees of freedom* for each molecule (six!).

For the sake of simplicity, we can restrict our evaluation to the interactions involved in *a pair of molecules*, by assuming that this interaction is *additive*; we can then consider what occurs in a population of *n* molecules. However, this is only an approximation, and we already know that for a small group of three molecules the results are not very good! Actually, the problems involved in such a calculation would be considerable, requiring thousands of hours of computation time on supercomputers.

Rather than considering the structure of a collection of water molecules by using totally *ab initio* methods, it is preferable to simulate the evolution over time of the positions in space of a finite number of molecules enclosed in an isolated space, whose energy and volume remain constant. This research lies in the domain of *molecular dynamics*. The effect on each molecule in a system of the forces resulting from the presence of other molecules is studied. Equations of motion are solved simultaneously, and the trajectories and velocities are analyzed to calculate the measurable properties, such as viscosity, thermal conductivity, speed of sound, diffusion coefficients, specific heat, isothermic compressibility, and dielectric constant.

Only a relatively reduced number of particles can be tested by means of this method. Here, the advances in knowledge are dependent on progress in computer science, and the arrival of "neural network" computers is eagerly awaited. For a collection of rigid spheres, the number of particles can be as high as 10,000, but for water, given its number of degrees of freedom, we must be satisfied with fewer than 2,000 molecules. They are enclosed in a central three-dimensional box flanked on each side by identical cells for which the calculations are repeated in an infinitely periodic manner. The initial positions are fixed; for example, a molecule is surrounded by a first, then a second layer of coordination as in the classical models. The interaction is then engaged and everything is set into motion.

An *effective pair potential*[*] is used. In the case of spherical particles, such as noble gas atoms, it is necessary to consider the weakly attractive forces known as van der Waals forces (named after the Dutch physicist Johannes D. van der Waals, 1837–1923). The corresponding potential is proportional to the sixth power of the reciprocal of the distance between the two members of

---

[*] *Effective*, because it is an assumed potential, with the hope that it is an appropriate representation of the interactions that we do not know how to describe in detail.

the pair. A repulsion term is added to make sure the particles do not collide with each other, which can occur (if we don't watch out) in a computer simulation, that is, of course, completely imaginary. This situation results in the Lennard-Jones potential, which for water is applied to pairs of oxygen atoms. Since the nonspherical shape of the molecule requires an unequal distribution of electric charges, it is necessary to add other terms. Many theoreticians now believe that no actual electron transfer occurs between molecules along the extremely flexible hydrogen bond. This bond is not a true chemical bond, like a covalent bond, but rather a strong electrostatic attraction that can be described by a Coulomb potential. The trick is to find an adequate model for this potential, starting with the assumed distribution of point electric charges in the water molecules, the polarizabilities, and the probable geometry of collisions (the known structure of the dimer assists greatly in making choices). The water molecule can even be dissociated so that $2n$ atoms of hydrogen and n atoms of oxygen are considered, which leads to finding the potentials of pairs when the molecules will form automatically. An enormous amount of computation time has been devoted to these problems over the past few years.

Several research teams have been working on this problem, including the IBM team directed by Enrico Clementi in Kingston, New York. This effort has culminated in a simulation of liquid water by means of a calculation method, known as the Monte Carlo method, that uses a sampling technique to generate successive geographic molecular configurations, each of which corresponds to a particular state of equilibrium for a system with n particles. This technique is interesting because it can be used to simulate several interesting properties of water.

Some scientists believe that these massive simulations, these "gargantuan," "brute force" techniques, used for a relatively small number of molecules, represent an excessive use of computation time, and that simple macroscopic models serve just as well in reproducing experimental results. The debate cannot be summed up as a disagreement between "rich" and "poor" scientists, because there is a huge problem: the simulations do not really confirm the classical models! Neither mixed microstates nor "icebergs" are found; in sum, none of the properties of those models that were so carefully worked out from the results of precise spectroscopy!

This conflict undoubtedly reflects not only the persistence of mental blocks related to the problem of representing structures, but also a refusal to lose the intuitive feeling for and the architectural image of chemical bonds in the computer's "crunching" of thousands of interactions in an exercise that is impossible for the human mind to follow. Yet today we are aware of the many problems in chemistry that result from the interplay of a large number of components without being able to decide which ones are most important. To simulate the precision of experimental measurements, there is no other way than to consider everything and to be resigned to the necessity of massive technology use. Thus the computer functions as a prosthesis for human understanding, which finds it difficult to deal with such complexity involving hundreds of variables!

The strongest image that can be found for the "structure" of water is provided by video simulations that represent the results of position calculations for each atom in each molecule for a total system lifetime that is just a few *picoseconds* (1 picosecond = 10^{-12} second). This technique reveals a world in motion, in which hydrogen bonds are constantly and rapidly being formed and broken, while the whole, continuously reconstructed network appears to pulse very slowly. Rapid rotations superimpose their rhythm on the slower background of translational motions. Thus, it appears that the groups of "five" of the "classical" model exist, but that their lifetime is extremely short. We cannot deny that the scientific knowledge of liquid water now depends on the subtlety of these dynamics simulation studies of molecular motion. These computer-based experiments offer a rare means of visualizing what happens on the molecular level between water and something else, such as the wall of a container, a large biological molecule (such as DNA), or a metal ion.

THE WATER OF LIMITS

To further complicate, as if we needed to, the question of ordinary water, there are fluids that are certainly water—

which we will call "neo-water," if we may—but which to all appearances are totally different from water in terms of physics and chemistry. Previously laboratory curiosities, these fluids are today candidates for sometimes spectacular industrial uses!

Supercritical water

Water is a solvent and a medium in which chemical reactions occur in nature as well as in industry. In general, increasing the temperature accelerates the reaction rate. We know that, in the kitchen, for example, boiling water is more efficient than warm water for cooking pasta, and using a pressure cooker, in which the temperature of water under pressure is slightly greater than 212°F [100°C], considerably reduces the cooking time for vegetables. Under certain circumstances, both the temperature and the pressure can be increased even further. In 1821, Baron Charles Cagniard de La Tour (1777–1859) discovered that there is a temperature beyond which a substance is neither a gas nor a liquid but a fluid that encompasses the two states: *supercritical fluid*. This fluid is formed beyond the point where the densities of the liquid and vapor under pressure become equal. Ordinarily, the density of a liquid decreases as a result of thermal expansion as the temperature increases. Water becomes a *supercritical* fluid when the temperature and the pressure exceed the critical values for temperature

($T_c = 374°C$) and pressure ($P_c = 221$ bar). At this point, the density of water is 0.3 g/cm^3.

We have seen that the physical properties of water depend particularly on the pressure and the temperature; some of them—the viscosity and the dielectric constant—follow more or less the evolution of the density. The same is true of solid bodies in water: the coefficients of diffusion for dissolved substances decrease as the density increases. The specific heat at constant pressure becomes infinite at the critical point, as does the thermal conductivity. The volume of the supercritical liquid can diminish considerably if a substance is dissolved in it, because the solvent molecules aggregate compactly around the "foreign" molecules. The static dielectric constant of ordinary water is high (78.5 at 25°C and 1 atmosphere), which explains why it easily dissolves ionic salts. The dielectric constant of supercritical water, however, is much lower (6 at the critical point), which allows it to dissolve nonpolar compounds, just like an organic solvent does.

The possibility of developing an entirely new chemistry with this fluid is still in the exploratory stages. For example, in only a few minutes, hazardous compounds, such as PCBs, can be oxidized and transformed into water, carbon dioxide, and inert substances which are then separated upon their return to normal conditions. Substances such as oxygen or methane are soluble in supercritical water: combustion can occur in the *presence* or *absence* of a flame! This could eliminate quite a

number of difficulties relating to high-temperature in-
cinerators, particularly the release of waste into the
atmosphere. A prototype is operating at 600°C and 250
bar, that is, under conditions at which water is not yet
corrosive for the walls of the container. Perhaps this will
pave the way for the clean disposal of many industrial
chemical wastes.

Supercooled water

It is generally believed that water freezes at 0°C, because,
by definition, ice *melts* at 0°C. However, water that is
carefully cooled (*supercooled*) can remain in the liquid
state at temperatures as low as –40°C. This phenomenon
also occurs in many other materials. For a solid to appear,
a *seed* of a crystal must be present. The appearance of
such a seed is a random event that can be triggered by the
presence of dust, a rough surface, and especially by
substances whose surfaces, on an atomic level, are structurally similar to the elementary crystalline unit cell of
ice. For example, molecules of certain alcohols form a
hexagonal linkage on the surface of water, which allows
it to freeze practically at its melting point. In light of this
finding, clouds can be seeded with substances that accel-
erate the formation of ice, the precursor of rain, which in
the atmosphere ordinarily appears only below –10°C. For
a number of years, other alcohols (glycols) have been
used in antifreeze to maintain water in the liquid state in
automobile radiators.

The French chemist Victor Regnault (1810–1878), who for a hundred years held the record for the preparation of supercooled water (–32.8°C in 1847), was able to measure certain of its physical properties. Water droplets 5 to 20 μm in diameter dispersed in a cloud, in the absence of contact with the walls of the container, can remain liquid up to –41°C. The density of this supercooled water decreases rapidly with temperature down to –45°C, when it reaches the value corresponding to the density of ice Ih. The specific heat of water increases exponentially below –10°C, which indicates that it is approaching a pseudocritical type transition: an "identity" between the liquid and the crystal, as is the case for liquid and gas described above. Supercooling is a phenomenon that may play a role in nature by contributing to the cold-resistance of plants or animals, such as crayfish.

Some bacteria produce proteins that can act as seeds for ice crystals. This is the case, for example, with genetically altered *Pseudomonas syringae*. Genetic engineering has been used to remove the gene coding for this kind of protein. This could be a boon for agriculture if, as is hoped, this modified bacteria can be introduced to diminish the frost danger for frost-sensitive crops, which are sometimes destroyed by a cold wave, in the Mediterranean or the Caribbean. But legislators are watchful: ecologists in California, claiming that genetically altered bacteria endangers the natural equilibrium of bacterial populations, have already brought suit to prevent the use of such bacteria. For commercial purposes, another approach was taken. Bacteria that produce substances causing water to freeze as close to 0°C as possible were introduced into the water used in snow machines! We do not know whether the variety of sources that can serve as seeds for crystallization explains the amazing diversity of snow flakes. These crystals have proven to be very difficult to reproduce in the laboratory, although it seems that scientists at the University of Hokkaido in Sapporo, Japan, have succeeded in doing just that. There are at least thirty different forms of snowflakes! The Eskimos have forty-six different words to describe different kinds of snow!

There is yet another strange type of water discovered by the French chemist Marcelin Berthelot around 1850. This water, subjected to negative pressure, has a density of 0.7 to 0.9 g/cm^3. It can be found in the natural state in the form of microscopic inclusions in minerals, such as quartz or fluorite. The melting of ice composed of this water occurs at 6.5°C and involves tensile strengths close to 100 megapascals (Mpa). This observation gives us some idea of the attractive forces of the intermolecular potential. In a fluid under pressure, this potential is predominant over the repulsive forces within a small radius. Under normal conditions, the repulsive forces determine the structure and the dynamics of liquids.

VI

WATER AND

OTHER THINGS

It is confession time: The previous two chapters paint a portrait of a water that exists only in the mind of the researcher. This portrait is derived from calculations, or it is embodied in the form of a temporary geometry deduced from the interpretation of spectral signals resulting from the passage of a collection of molecules through a device. Actual liquid water, "our" water, is a society of molecules that has several thousand members and occupies a space bounded by the walls of the container on the one hand and by the horizontal boundary subjected to the pressure of another fluid—air—on the other. There is nothing pure about this population. Whether we like it or not, there is always something else in water, such as atomic remnants snatched from the solid surfaces of the container or the chemical substances resulting from the dissociation of water itself. In any given volume of water, in addition to water molecules, there is a small number of ions. Simple ions are formed from a single atom; complex ions are negatively charged molecular groups (anions) or positively charged groups (cations). Water can dissociate into hydroxyl ions (OH^-) and protons (H^+) that immediately are solvated to become hydronium ions (H_3O^+). The molar concentrations of H_3O^+ and OH^- are quite low:

each is only 10^{-7} at 25°C. Chemists use the symbol *pH* to represent the negative logarithm of the hydronium-ion concentration. The pH of "normal" water is 7. The lower the pH value, the higher the concentration of hydrated protons, and the more acidic the medium. Conversely, if the OH^- concentration is greater than 10^{-7}, then the pH is higher than 7, and the medium is basic. Basic or acidic aqueous solutions result from the addition of bases or acids to water.

Water, which poets portray as being so gentle, is actually an extraordinarily aggressive substance. It can carry and disperse a great many chemicals that can then be deposited or abandoned wherever the water may take them, that is, where the conditions favoring their solubility disappear. This property causes problems for not only the legislator, but also for the chemist. Water insinuates itself everywhere: it is found in clouds and in clay, as well as in the human body. It carries many different things from one point to another on the globe. Water is the force behind erosion, it can erase traces, scatter food, debris, waste; it can escort them, distribute them, bring them to places where they can sometimes have a lethal effect. Water gathers together, it concentrates. When rain from a storm falls after a long drought, the water carries toxic substances, deposited in drainage channels by industrial pollution, into the sewer system. These toxic substances then enter rivers, where they can poison fish. Some of the dead fish end up on river banks; the rest, carrying the poisons that killed them, are trapped by the screens and filters of water purification plants. Since it is impossible to describe here all the components involved in the mechanisms of transport, dissolution and/or deposition of salts, molecules, colloids, and polymers, and because the scale involved can be as large as a continent or as small as a tiny vessel connected to the membrane of a living cell, we will instead discuss some examples.

WATER AND SALTS

Water easily accommodates positive and negative ions which, in the solid state, electrostatically combine to form salts. Sodium chloride forms Cl^- and Na^+ ions when it is dissolved in water. Salts are easily dissolved by water because the high value of its dielectric constant allows the opposite charges of the ions to be separated. However, there is a limit to the concentration of each ion in water, because each salt has a characteristic constant number, known as the *solubility product*, at a given temperature and pressure. If the product of the molar concentrations of the salt ions is equal to this value, then the salt *precipitates*. The small seeds of solid crystals which form spontaneously in the liquid grow until the product of these ion concentrations returns to a value below that of the solubility product. As these products are quite different for each salt and vary by several orders of magnitude, salts can be separated from each other in the solid form by fractional crystallization that takes advantage of concentration differences. This, for example, is the method used to obtain salt from salt marshes.

Some salts have extremely low solubility products, which means that they do not readily dissolve. If a large amount of a salt is dissolved, it can easily precipitate as a result of changes in temperature or pressure that affect the value of the solubility product. This is the case with calcium carbonate ($CaCO_3$), the principal component of natural limestone rock. Limestone will dissolve in water containing carbon dioxide, which is readily absorbed by water up to a maximum of 878 cm^3 of CO_2 per liter at 20°C. The water then becomes acid, and chemical equilibrium favors the activity of the bicarbonate ion (HCO_3^-), whose calcium salt is more soluble. However, if the amount of dissolved CO_2 decreases, for example by aeration of confined subterranean water, the calcium precipitates because the HCO_3^- ions react to form CO_3^{2-} ions, which complex with the calcium ions (Ca^{2+}). This process occurs in caves, where impressive stalactites and stalagmites are formed. The same mechanism is at work in the more prosaic deposition of scale in water pipes, which is nevertheless actually quite complicated. Often natural water is supersaturated with calcium carbonate, which does not precipitate. For the precipitate to cause trouble, it must cause encrustation, which results in compact and adherent deposits. Although such phenomena cannot be predicted by thermodynamics, they nevertheless have considerable practical significance. To cite only one example: scaling in pipes carrying coolant limits the power supplied by some nuclear plants.

In theory, therefore, a supersaturated solution precipitates in the presence of a crystal seed, although the dynamics of the solid's appearance can be slowed or inhibited. Stable saline solutions that have largely exceeded the equilibrium of solubility can thus form, but they are likely to liberate the salt they contain all at once, which is what occurs, for example, at great depths in the Red Sea. This may also explain the existence of huge deposits of sea salt at the bottom of the Mediterranean, which has led some geologists to speculate that this sea was once entirely dry during a recent epoch.

Water can thus contain ions, dissolved gases (19 cm^3 of air per liter of water at room temperature and pressure), and molecules in rather variable amounts that can easily be modified within defined limits by changing the temperature and the pressure. Some salts, known as hydrated salts, separate from water by crystallization and carry along with them a specific number of water molecules that constitute an integral part of the structure of the crystal and its chemical formula. These water molecules are often "stuck" to a cation and form a regular geometric polyhedron around it; the solid then incorporates these units. Mineralogy provides many natural examples of this phenomenon.

The structure of crystals containing water strongly bound to cations has led chemists to speculate on the importance of the local structure of water around dissolved anions and cations in solution. The water molecule shields separated electric charges. We can imagine that a charged species, such as an ion, seriously disturbs the formation of relationships that occur by means of hydrogen bonding among water molecules. In fact, the presence of the ion is like a disturbance caused by a stone thrown into a peaceful pond—existing bonds are broken and other preferential bonds appear that are directed toward the intruder. If the charge on the intruding ion is sufficiently large—for example, if it is a trivalent cation—it can polarize water molecules over large distances; consequently, they all align with their negative poles pointing toward the cation. In this way, a small "iceberg" is formed in which a geometric structure can sometimes be detected: it is a symmetrical, regular arrangement of water molecules. A similar phenomenon occurs with anions (in this case, the positive poles of the water molecule are aligned to point toward the anion) or any other charged surface, such as the wall of a container, a small particle of solid in suspension or a large biological molecule. When a *nonpolar* molecule, which has an equal distribution of electric charge, is dispersed in water it embeds itself in the interstices of the "structure" of water. The water molecules retain the hydrogen bonding among themselves and surround the "intruder," which is thus imprisoned in a sort of cage with a more or less regular geometry. A structure of this type in which molecules of a nonpolar substance are completely enclosed within molecules of a polar substance is known as a *clathrate* compound.

The methods used to study these associations between water molecules and charged or uncharged species are the same as those described for pure water. X-ray and neutron diffraction or vibrational spectroscopy are used to determine the number of water molecules surrounding a cation or an anion. Computer simulations are also used to provide some idea of the molecular dynamics around the charge. Studies of this kind, which are tremendous in number, give rise to the same disagreements that the models for the "structure" of "pure" water do. To cite but one example, a video simulation of molecular dynamics produced in an IBM laboratory shows a lithium formate solution. The cation (Li^+) is shown being tossed about in a cage formed by the first layer of six water molecules which slowly exchange (over a period of several picoseconds) with the thirteen molecules in the second layer. The formate anion ($HCOO^-$), which is much larger than the lithium ion, is much calmer, moving less, and is surrounded by twelve water molecules, each with a pair of hydrogen atoms directed toward it. The water molecules of the first layer sometimes exchange with thirty-six water molecules in the second hydration layer. All the molecules move and vibrate, turning upside down and slowly making translational shifts.

WATER AND LARGE MOLECULES: LIFE

Proteins, peptides, and nucleic acids—to name but a few—are large molecules involved in the chemistry of life. Since all living entities contain water, it is of paramount interest to discover the relationship between these gigantic entities and the tiny water molecules. Water acts primarily as a solvent, a medium in which the chemical reactions forming the basis of biological processes occur.

A protein or a complex molecule, such as DNA, has many parts. Since it occupies a considerable volume, it has an extensive external envelope. Some areas of the surface are *hydrophilic*, whereas other, uncharged, areas are *hydrophobic*: water can organize around them in "clathrate" structures, and these zones can also reinforce cooperation between solvent molecules by "separating" them from contact with the nonpolar zones.

When a charged hydrophilic surface comes into contact with a fluid, a local order can form as a result of the polarization of the water molecules, which can extend over a relatively large depth (3 nm). Often, as a result, the freezing point of the solvent is lowered. The skin of certain fish that live in very cold waters contains proteins of this type, which inhibit ice formation.

Protein mechanics are complex because protein molecules fold and unfold. Their reactivity depends to a large extent on their spatial form, and thus on their stereochemical or chiral properties. The cohesion of the atomic skeleton and the distribution of electrical charges at the surface are also important parameters. On the average, 55 to 75% of the surface is hydrophobic, 15 to 40% is polarized, and 5 to 20% is directly charged (ionic). Water plays a role in the stabilization of structures, particularly through the architecture of hydrogen bonds. Sometimes the hydrophobic sites fold over on themselves inside the protein, protecting essential chemical groups in the cavities from which the solvent (water) is thus excluded. Connections between molecules occur just as frequently through their hydrophobic groups.

In May 1992, it was discovered that water plays an even more important role than previously thought in the basic chemical reactions involved in the mechanisms of life. For example, when four oxygen atoms bind to hemoglobin, *sixty* water molecules are also bound! When such significant amounts are fully involved in a chemical reaction, they affect the thermodynamics of the system. The contributory mechanisms of water fixation and loss on the surfaces of large molecules probably also play a decisive role in many essential reactions. These discoveries were made by using new experimental methods, such as sensitive measurements of osmotic pressure. For instance, when glucose binds to an enzyme, the enzyme releases around one hundred water molecules. Thus it is clear that the role of water in biochemical reactions is not limited to its action as a mere solvent. Water is a full-fledged ligand that is directly involved in chemical exchanges, and may even play a key role in these processes. For example, to open a channel in a biological membrane, at least a thousand water molecules must congregate in the immediate vicinity of the portal of entry! For a reaction to take place between two molecules, it is clear that the water molecules present on their surfaces must first be eliminated or displaced. This is a very essential step in the reaction. Then the water, acting as a ligand, can directly assist in the associations formed. Once its work is done, it can again surround the collection of molecules. The cycle of water in biological reactions can no longer be ignored, inasmuch as the protons circulating in the water transfer electric information across membranes.

The dynamic mechanical relationships between different parts of the protein and the solvent are just one of the elements of the complexity involved in the series of reactions that put into action the actors of the components of different molecular structures. Embedded in these components are metal ions as well as active chemical groups containing sulfur or phosphorus, or even classic molecules such as sugars.

An essential research goal consists in determining the positions of water molecules that are structurally associated with proteins, either those directly involved in the structure (as a part of the ligand bound to a metal ion, for example) or those involved as surface components attached to the hydrophilic portions, or again as structural elements of the solvent in the vicinity of the molecule, or lastly those included in the "bridges" between different molecules. It is difficult to obtain experimental data on these different situations. X-rays are used for this purpose, but we know that they detect only the oxygen atom in the water molecule. The intensity of x-rays from synchrotron radiation has nevertheless made considerable progress possible in our understanding of protein structures. The Brookhaven Protein Data Bank in New York provides crystallographic information on about 30,000 oxygen atoms in the water molecules of proteins.

Another recent technique that yields interesting results is *nuclear magnetic resonance* (NMR), which is based on the excited states that occur in certain nuclei. The nucleus of the hydrogen atom, a proton, has states that are normally "degenerate" (they have the same energy), but application of a magnetic field results in the creation of an excited state, because a very small energy gap appears that is located in the radio-wave region. Radio waves having frequencies corresponding exactly to the energy gap will be absorbed by the nucleus in the ground state. Using modern electronics, this frequency can be measured with great precision. In fact, each proton in a molecule can be associated with a particular resonance frequency that depends on its chemical environment. NMR is a conformational analysis method that is extensively used by chemists to determine the structure of molecules. It is easy to make the protons in water resonate (at 300 megahertz [MHz] for a field of 7 teslas), and this property is used to construct "proton magnetometers" which use resonance to measure inversely the fluctuations in the Earth's magnetic field.

The most important application of NMR is *medical imaging*. When a magnetic field varies in space, differences in the distribution of water throughout the body appear as contrasting images that reveal the contours of organs. Depending on the part of the body, the images, which have a very high resolution (on the order of millimeters), represent strong or weak absorption of radio waves. For instance, the location of a tumor can be determined by focusing on biological targets molecules containing a metal cation that produces a magnetic perturbation, thus producing specific contrasting images.

For proteins, it is usually difficult to distinguish the NMR signal produced by water bound to a molecule from the signal produced by other protons in the same molecule. All possibilities of modern electronic techniques are enlisted to manipulate the excitation of a system of protons to follow its relaxation in controlled-time stages. In this way, coupling between neighboring protons can be evaluated and the residence time of water molecules lodged inside or glued to the surface of proteins can be measured. Water molecules remain a "long" time (on the order of 10^{-2} to 10^{-8} second) inside proteins, and a very short time (less than 1 nanosecond) on the outside of proteins. It is as though there were a row of soldiers, lined up at attention at the base of a wall, facing a crowd of disorderly troops at ease. Individuals from one group are constantly and rapidly trading places with those in the other group. At no time does the ordered, well-organized line against the wall ever disappear. However, if the protein is removed from the system, the process naturally ceases, and the sentinel water molecules on the surface disperse immediately because the electric field that maintains their alignment no longer exists.

Computer simulation produces a picture that closely resembles this description: water molecules are seen crowding around electrically polarized areas or around anionic or cationic groups, which, assuming well-defined orientations, sometimes connect different parts of giant molecules by chains of hydrogen bonds, with each molecule showing a tendency to jump constantly from one position to another. Simulation also demonstrates that certain processes can occur only with the active cooperation of water molecules. Thus, for a potassium ion to cross a membrane through an open channel in a peptide structure, it needs the "help" of two water molecules that appear to push and pull it through.

Relationships between water and biological matter have been the subject of intense studies, which are facilitated by sources of more and more powerful x-ray sources, such as those of the European Synchrotron Research Facility (ESRF) in Grenoble. These new radiation sources allow the structures of proteins to be determined in time sequences covering the reaction steps. In this way, we will learn more about the role of water in the physicochemical phenomena involved in maintaining life.

WATER AND EARTH

Silicates make up the largest part of the Earth's crust. The silicon ion (Si^{4+}) is surrounded by four oxygen atoms that are arranged in a tetrahedron to form SiO_4^{4-} anions. They can also condense to form (Si_2O_7) pairs, rings, or infinitely thin films of $(Si_4O_{10})_n^{4n-}$. The electric charge of these groups is balanced by cations. The most abundant cation found in nature is trivalent aluminum (Al^{3+}), but magnesium (Mg^{2+}) and potassium (K^+) also play important roles in minerals. Water and the ions derived from it, particularly OH^-, participate in the formulation of these complex structures, especially those of the silicates that present a layered structure (clays). The minerals of clays, such as micas, all have complex and variable chemical compositions; they are easily cleaved along parallel layers. These layers are often held together by cations coordinated with water, as is the case with magnesium and aluminum. In other cases, such as with talc and pyrophyllite, since the total electric charge is balanced, water is repelled by the hydrophobic mineral, which it cannot cause to swell. The kaolin group contains minerals such as montmorillonite that has cleavage faces with exposed hydrophilic OH^- groups. They readily form gels with water by increasing the hydrogen bonds between the solid and the liquid.

The *clay fraction* contained in soil is defined as the mass of particles smaller than 2 μm. These small hydrophobic or hydrophilic crystallites are easily dispersed in water. When the concentration of the solid in this slush is very high, the viscosity of the fluid increases until a plastic paste is formed. When this paste has a consistency somewhere between that of soup and soft soap, it is known as *mud*. Both mud and a ball of clay that has been gradually softened by kneading are in an intermediate state in which water mixes very intimately with the solid. The material escapes, on the one hand, from the brittle rigidity of the crystal and, on the other hand, from the disappearance of water that the hand cannot hold. Thus, forms can take shape beneath the fingers: water fragments the small crystals, while at the same time playing the role of a lubricant that allows the crystalline faces to slide over each other; it holds them together. Once this work is done, heat from fire first evaporates the water, then, as the temperature increases, the structural hydroxides are eliminated from the clay, finally solidifying the shape given to the soft material by transforming it into a stable silicate on which water no longer has any effect.

As surprising as this may seem, mud or modeling clay, one of the first materials used by humans to fashion objects, is a relatively new subject of scientific study. Only now is it being analyzed as a complex fluid material endowed with controllable mechanical, chemical, and physical properties. Mud is now considered to be like a disordered medium that is capable of being analyzed by statistical physics. As a fluid, it is characterized by a non-constant viscosity that depends on the strength and the rate of the stress applied: mud can fracture like a solid. As a material in contact with others under circumstances where the mechanics of the mixture depends on natural conditions (diffusion, percolation, porosity), it forces us to resort to such sophisticated mathematical tools as the theory of fractals to understand the forms observed. In this case, mud has an unexpected importance, because it can serve as a model that researchers can manipulate to explain phenomena involving other viscous fluids on a geological time scale: the flow of volcanic lava or the stages in the gestation of liquid hydrocarbon deposits. The practical importance of research on mud is that, as an industrial raw material, it is used as a lubricant in drilling oil wells, or as a stopgap in civil engineering. The techniques used to study mud can also be used to study adhesives or gels, or any viscous states that can easily accommodate various materials which can then be permanently incorporated following heat or chemical treatment. This is the case, for instance, with cob walls made from dried mud and straw, a primitive example of a composite material.

VII

MYTHICAL WATER

The instruments used for chemical, physical, or statistical analysis are just prostheses. At the tip of their electrodes or through the prismatic window of their spectroscopes they gather data associated with entities that escape direct perception. Science uses these measurements to describe matter as a society of atoms or of molecules in stationary rows or in dynamic action. The relative numerical proportions of armies of chemicals combined with the state variables of temperature and pressure determine the appearance or the future of matter. To understand and learn more, we need to make Gulliver's voyage through this Lilliputian universe. We do this by interpreting images on the paper, the screen, or the photographic film that are the result of interaction of an exploratory probe with the infinite swarming of the entities which science has created from mathematical virtues, but which no one has ever been able to hold in the palm of the hand as isolated, tangible objects. Today, to believe in science is to believe in the power of the image as a representation of reality. Despite this illusion, at the beginning of the 17th century, inspired by the battle cry "our senses deceive us," Descartes, Galileo, and others paved the way for humanity to use images for the purpose of prediction, something that was never accomplished by Religions, Philosophies, or the magi associated with the Correspondences.

Today, science is going even further, always inventing new images, new models, improving the musings embodied in the tangible and intangible, so many levers that lie heavy on society. The domesticated photon and electron are inaccessible abstract entities, objects that exist only in the human imagination, yet their manipulation allows us to act directly on matter and to change the course of everyday life. Science also gives nations power. Scientific laws, with their arsenal of mathematical formulas, appear to be like elements of concrete action: they are transformed into machines that extend the capabilities of our physiology, experienced by our bodies, observed by our eyes, or explored by our fingers; familiar sensations, within our cultural frame of reference. But as basic science increasingly distances itself from the immediate experiences of the senses, it becomes more and more difficult to explain to those outside a small circle of specialists.

The result is that the modes of ancient knowledge based on the functions of the body or the magical beliefs of paganism derived from spontaneous animism, only slightly influenced by the good Book, are now making a comeback. They are obviously the crucible of a great deliverance, as if the possibility of knowing, once again, immediately, by contact, by the laying-on of hands, by personal experience, would open vistas that had previously been closed under the yoke of the dry studies required. This step backward, this regression toward the irrational, allows the old theories to get back into the saddle. For example, we see on the political (ecological), commercial, and artistic horizons a return of the traditional division of matter into Four Elements, a division that was never entirely banished from the unconscious.

These rediscovered Four Elements are just the foundation for the return to Nature, packaged in the raiment of the Environment. The Goddess is once again among us; she no longer has to rely on the specialized, sometimes cruel, explosion of Polytheism. There she is, virtuous, righteous, her altars ready for devotees. It is not polite to note that the immense effort of science over the past four centuries has done little more than assist the never-ending inclination of humanity to replace Nature with something better, more "efficient," artificially high-powered. This is as true for the art of gardening as it is for plastics. This remark can outrage only those who have always been shielded by the comforts created by science and technology so we can deal with cold, dirt, hunger, thirst, and fatigue. The feeling for Nature today, as it was in the past, is a nostalgia for riches, a fantasy book of safaris, a dream of an Eden without mosquitos or snakes.

In the early days of contemporary science, at the time of the first models, the tangible, visible properties of water were largely used by scholars to explain a variety of things. The capacity of water to be absorbed by some solids seemed to be a natural process in the passage from one elemental state to another, and the sponge model, or the property of sponginess, has been used to explain meteorological phenomena (the air as a "sponge" for water) as well as heat exchanges (via "igneous" fluid). Other metaphors related to devices using water, such as the pump and the filter, fed the first explanations that now seem naive. Nevertheless, they manifest the desire to understand these phenomena by comparison with direct observation of the way the Four Elements combine with each other.

In the Gang of Four, Water plays a special role because it is everywhere. Combined with Earth, it makes mud; kneaded with clay or with flour, it transforms what is hard and dispersed into a soft mass that can be deformed and shaped, then become rigid again when acted on by Fire. Water is an agent of transformation that the human hand puts into action. Earth receives Water and blossoms. Is water a masculine principle or is its softness, spread over voluptuous and penetrable surfaces, the affirmation of a femininity ready to surround and/or smother? Water can be heated and cooled easily; scalding hot, it is the agent of Fire whose energy is transmitted to the contents of the cooking pot where, again, it assists in a transformation. Icy or cool, it soothes the heat imposed on the body. If Fire is the source of Light, then everyone can see that Water allows Light to pass through its transparency, while at the same time reflecting it, shifting it, capturing it to finally scatter it in space. And if Fire becomes too violent, Water quenches it. Ally, ruler, or medium of transportation? And now as to Air, isn't it obvious that these two Elements blend together?

See how vapor rises in the Air, how the clouds float, how Water itself so easily dissolves gases to form sparkling bubbles! Air and Water are totally wedded, as the Fog attests. Water is the Companion, the Spouse. Its field of interactions with its Fellow Elements is more developed than among any of the others, which fact is the source of its domination in the primary neopanoply of the Goddess Nature.

We are indebted to the Ancients for another bit of theology, the subject of the dispute about the universals: Do Beauty, Good, Truth, and the Human Being really exist? We have known since the time of William of Ockham (ca. 1285–ca. 1349) that what cannot be touched with the hand does not have a real existence. Today, for water—as for air, and to a lesser degree for earth and not at all for fire—the dispute about the universals focuses on the Pure: that which seems uncontaminated, untarnished, possessing all its original qualities, as it did during the first few hours of the World. This strong notion, so solidly established, is totally in conflict with the notion of the Pure as it is understood by the recent center of scientific knowledge: in this context, purity is associated with a number; it is never absolute, but always relative. Nevertheless, in an almost religious way, the myth of the Pure has been built into and surrounded by another myth, the myth of Nature, which has major commercial and technical consequences!

We have convincingly argued in the preceding chapters that water is a serious and complex subject of scientific investigation. A great number of original articles on the subject have been written, and the compilations describing and critiquing the results fill dozens of thick volumes. Hardly a month passes without some new information on water appearing in the scientific literature, as has occurred on several occasions during the writing of this book. Thus water is not a "dry," tired subject; on the contrary, it is the subject of spirited, high-tech research.

We are becoming more and more aware of the hard, substantial content of science, developed within a very small community in the communion of a cryptic and secret language, which is not released to the outside world, to society, unless it is carried by the recycled cast-off of a myth. In this transmission process, scientific material is very often distorted, filtered, transformed, and sometimes misrepresented. So we have the Big Bang and the origin myth; the ozone hole and the greenhouse effect with the Cosmic Egg (which is in danger of being pierced or rotting from the inside), the image of radioactivity (characteristic of the Genie, beneficial when imprisoned in the bottle, harmful when allowed to escape), or even science *at large* as an activity worthy of the Sorcerer's Apprentice.

Water is a composite of powerful myths, often risqué (such as the symbol of the interplay of clouds and rain among the Chinese), because they tend to be related to sexuality and reproduction. The water that falls from the sky, the sperm of the gods, fertilizes the soil, the

Mother Earth goddess. But water is also, in the bosom of the sea or at the edge of a spring, where feminine, dominating, self-confident divinities or timid virgins live and hide.

This idea that water embraces femininity, that it carries something marked with the sacred seal of reproduction, that is, of temporal continuity and of renewal, confers upon it a magical character, especially if it lends legitimacy to places consecrated by tradition or by custom of miracle or rebirth. Thus, all over the world are places where there are marvelous waters which devout pilgrims visit, carefully taking home some in bottles as a beneficial talisman. In this case, water has the quality of an intermediary between the sacred donor, which, in its goodness, yields to the liquid a fraction of its Being, and a chosen, deserving recipient, who humbly receives the mark of the beneficent communion with the Supreme. This is, of course, the definition of *Holy Water*. Its use does not necessarily correspond to the current thematic of the sexual act, nevertheless there is something in it akin to impregnation.

Aside from the influences of the religious and the mythical, there is another possible source of contamination in the transmission of scientific results. It is the clandestine embarkation on the ship of the productions that formed the basis of knowledge before the appearance of contemporary science at the beginning of the 17th century: the Correspondences. The Correspondences are universal; they touch on everything, as shown so well by Michel Foucault in the first chapters of his book, *The Order of Things*: they unite stars with plants, plants with metals, metals with parts of the body, parts of the body with planets, etc. It is like a long serpent that bites its tail, without the nature of the connection between such different "things" ever being specified other than as the result of a mysterious action at a distance in space or time whose signatures the adept must seek everywhere in Nature.

The series of Correspondences thus appears to be as firmly entrenched as ever, hardly affected by four centuries of scientific research. Its themes are largely and complacently carried by the media, from television to the publishers of pseudo-scientific books. It is difficult for the public to tell the difference between the output of faithful adherents of the Correspondences and that of research laboratories. Even worse, ambiguity occurs: voluntary or clumsy compromises appear in scientific output, and branches of the art of Correspondences—as old as astrology or as new as homeopathy—cover themselves with the mantle of science by aping its language and calculations, and even the appearance of the experimental method, thus attempting a sort of forced entry into the world of science.

Water is the object of choice on which both the strength of myths and the seductive allegorical pleasure of the Correspondences converge. It is easy to rush into the breach opened by the complexity of the physical properties of water, the large number of factors involved, and the gaps in our knowledge of water. Seemingly rational detailed explanations are offered which are almost modestly expressed—with such hedging as "maybe" and "it seems that everything occurs as if…"—but which actually have the great expectation of providing an experimental scientific basis for the beliefs associated with the myths or the allegations of the Correspondences.

FROM THE "PICCARDI EFFECT" TO THE "MEMORY OF WATER"

The case known as the "memory of water" began well before publication of the famous article by Dr. Jacques Benveniste and his colleagues in the journal *Nature* on June 30, 1988. The likely inventor of the "memory" of water is the Italian professor Giorgio Piccardi of the University of Florence. He had been working since 1935 on the encrusted deposits of calcium in boilers which, as we discussed in Chapter VI, do not follow the predictions of thermodynamics. In October 1950, during a conversation with a Belgian researcher, Carmen Capel-Boute of the University of Brussels, who worked on the same subject, he brought up the possibility of a "memory" for water acquired after it undergoes physical treatment "involving only low energies." Decoded, this means treating the water either by radiation with electromagnetic waves in the radio wave range or by magnetic fields produced by magnets. We know that the structure of the crystals in a precipitate can be influenced by the presence of a magnetic field. The initiation of nucleation often depends on initial conditions that are unforeseeable, and a "helping hand" can assist in one direction or another.

Piccardi developed a specific experimental method for measuring these effects, using the hydrolysis of water to test an acid solution of bismuth chloride. The amount of time necessary for bismuth oxychloride to precipitate is recorded. Later on, we will see what to think about this method. Piccardi discovered that the sedimentation time depends on the treatment undergone by the water, not at the time of the reaction, but *several days or hours before-hand*. Over the years, the "Piccardi effect" would be used as an indicator for phenomena such as the presence of low electromagnetic frequencies, phases of solar activity, the electric potential of the atmosphere and, more generally, "influences related to the environment." Piccardi was more inclined to think of the phenomenon as a means of detecting the random effect of ambient cosmic rays on ordinary chemistry. But the key point, according to him, is that water functions as a tangible receptacle which undergoes and then retains the structural modifications induced by the presence of a weak external perturbation. The chemical reaction with bismuth just reveals the state of this "activated water." The measurements demonstrate cycles, sometimes with long periods (hundreds of days), which give these experiments their cosmic dimension, as if the invisible pulsation of the world were being measured. Piccardi died on December 22, 1972.

His ideas were not too well received by the scientific community, but the torch was passed, and studies of this type were carried out recently at the Center for Medical Bioclimatology Research at the University of Milan. The nature of the phenomena changed, involving only low energies, such as, for example, the "biomagnetic" energy emitted by the hands of "magnetic healers" or it involved the water from miraculous springs, especially shrines to the Virgin Mary (like the water from Lourdes). A new detection method was developed: the precipitation of colloidal gold by the hydrolysis of gold trichloride. The color difference of the colloidal solution (the "sol") is evaluated, according to whether the water is "activated" or not. If the water is "activated," it turns a more intense shade of blue. Later on we will discuss this type of measurement. There is, however, in these recent experiments a new fact: it is no longer necessary to carry the activated water from the sacred source. All that is required is to add three drops of miraculous water to ordinary water, and it will immediately acquire the desired properties: "Three drops of water bearing the original vibrations suffice to transfer the information. The new sample will have captured the particular frequencies of the original water." Even better, it can now be diluted as much as desired. The more the sample containing the three drops is diluted, the better it will work, especially if the precaution is taken of waiting at least twenty-four

147

hours between making successive dilutions. As for double-distilled water from Lourdes, it is superactive!

In France, the words "memory of water" were probably heard for the first time in public in May 1985, during a colloquium of the International Homeopathic League in Lyons. Two homeopaths from the French city of Montpellier, the Drs. Luu, developed the hypothesis that water is capable of "storing energy or, more generally, information, and restoring it on demand." They based their hypothesis on a variety of "information": on the one hand, a rather good knowledge of the different theories on the structure of water, and, on the other, a mystical act of faith, because, "according to tradition, Water, the elixir of life, must have a structure that allows it to play the role of a medium capable of trapping energy and of ensuring its transfer to living systems to provide the Heaven-Earth interaction." Consequently, "water is said to have retained the memory of having been touched by that which occurred first."

These closely related theories are derived from the basic concept that water has a structure that can easily be altered by external influence. Despite this structural plasticity, once a modification is acquired, it is conserved (which is slightly contradictory with regard to the first theory), until finally it is capable of acting on a suitable receiver. This is a thinly disguised search for a "scientific" principle that could confirm the existence of the connection between different objects separated in space that is assumed by the adherents of the Correspondences.

The work of the French researcher Jacques Ben-veniste provides a new link in this chain through the use of a biochemical reaction: the degranulation of human basophils triggered by very dilute solutions of antiserum directed against immunoglobulin E (IgE). Degranulation is measured by the microscopic examination of cells after they are stained with a reagent which attaches only to basophils that are *not* degranulated. The reaction is observed even if the initial solution, an extremely complex one containing many different chemicals, is highly diluted many times with the same liquid containing no antiserum. Better yet, the positive response that is observed with the dilutions shows a more or less regular periodicity. The authors were amazed, and noted that since "the original molecules are no longer present, the specific information must have been transmitted during the dilution and mixing process. Water could act like a 'ghost' of the molecule, for instance, through an infinite network of hydrogen bonds, or by electric and magnetic fields."

We know that an uproar was caused by the reporting of these experiments in a respected scientific journal. The editor of *Nature* was careful to state in a box accompanying the article that "there is no physical basis for such activity." The expression "memory of water" achieved a great deal of notoriety and even appeared on the front page of the French newspaper *Le Monde*, while outside observers monitoring the experiments in the Paris laboratory checked for fraudulent research practices, including sleight of hand. Nevertheless, unlike the case with the chemistry experiments described above, one can hardly speak of water as being involved as a primary component in the case of anti-IgE antiserum solutions. These extremely complex solutions contain many large molecules, salts, and sugars. Water is almost an incidental component, and it is unreasonable to seek an explanation for the observations that is based solely on a property attributed to water.

The affair of the "memory of water" and its predecessors have several common factors, which we should now discuss.

These works overlap with the classical myth of water as the favorite vector of the divine, which ensures a phenomenal success when public opinion intersects the circle of the collective unconscious, as long as the media are interested in the subject and the authors have social credibility.

The appearance of these works is indistinguishable from that of genuine scientific works. Often written by genuine scholars, they have all the aura of scientific texts: the language, the formulas, the illustrations, the data, the bibliography. Moreover, they contain sections that are completely normal. The *initial* articles often appear in prestigious scientific journals. However, *in all cases*, the results are based on an imprecise experimental method, which, in the last step of the process, requires the eye of an observer, with the inherent risk of subjectivity.

Finally, disproportionately inflated theories having semicosmological consequences are constructed on the basis of these results. The implication is that the experimental basis has been definitely established, or better yet, that it cannot be disputed.

The authors are obviously seduced by the boldness of their own theories, and their good faith cannot be doubted. However, the apparent convergences with social or commercial interests that can profit from the use of such propaganda for the pseudosciences appears questionable, especially if the work of respectable scholars can be brandished in the face of public opinion.

THE PATHOLOGICAL SCIENCE OF LANGMUIR

What happens to make serious scholars stray in this way? This is a social phenomenon, relatively rare in the scientific community, that nevertheless occurs regularly: outbreaks of *pathological science*. The physicist Irving Langmuir (1881–1957) described this phenomenon in a little-known text during a conference at General Electric on December 18, 1953. The U.S. government had asked Langmuir to investigate flying saucers, which led him to investigate the detection limit for physiological perception and to study a group of cases of famous scientific errors, spectacular phenomena discovered between 1900 and 1940 by renowned, established scholars. Later on, it

151

was shown that these phenomena did not exist, although numerous researchers had devoted a great deal of time and even many publications to the subject! Langmuir defined six criteria which would allow us to identify or suspect cases of pathological science.

Rule 1: "The maximum effect produced is due to a cause having a nearly imperceptible intensity: the effect is independent of the intensity of the cause." This is what occurred in the case of the dilutions described above. *Rule 2*: "The observations are always very close to the limit of visual detection, thus many measurements are required and it is hoped that some sense can be made from the data through the use of statistical methods." In the Piccardi effect, the sedimentation rate is measured by noting the height of the solid deposit a certain time after mixing occurs. The difference between the samples containing the control water and those containing the "activated" water is less than one-half millimeter for containers whose shape is not specified, nor is the technique for reading the levels described. As for the gold "sols" in the experiments at the University of Milan, the observer must use the naked eye to decide whether the intensity of blue is greater in one sample than another, and it is clearly stated that a spectrophotometer is useless because "the measurements must not be considered as lending themselves to colorimetry," since the absorption curves are always nearly identical! Finally, in the case of basophil degranulation, the microscope operator must recognize the difference between a pink-violet and a violet-pink color to tell which basophils have resisted degranulation. Thus, in all these cases, the experimental technique is based on a subjective physiological evaluation for a marginal phenomenon with a low signal-to-noise ratio.

Langmuir stated four additional rules concerning the social behavior of the actors and the structure of their discourse, including the penchant for fanciful theories that challenge classical science and an extreme personal tendency to always have a response to criticism and a particular delight in precise figures. Finally, Langmuir noted that although these affairs inflame scientific opinion for a short time, sometimes arousing animated debate, after a certain time they are relegated to almost complete oblivion.

The history of research on water experienced another "affair" of a slightly different nature. In 1965, the Russian scientist Deryaguin discovered "polymerized water," a type of water formed in fine quartz capillary tubes. While it has a lower vapor pressure than ordinary water, its viscosity and index of refraction are higher. It was believed that this was a type of polymerized water, and it was even feared that if this gel were to escape from the laboratory, it would contaminate ordinary water and replace it, which would be disastrous for life as we know it. Many researchers, including theoreticians from all over the world, worked enthusiastically on this fashionable problem under the spotlight of the media—the subject was even covered by the *Wall Street*

Journal—until August 1973, when the discoverer himself announced that "polymerized water" was just a solution of the silica from the quartz capillary tube.

Sometimes the scientific community likes to chase rainbows. However, in the case of the memory of water, there was great caution because it is obvious to all chemists that the experiments on the effects of highly diluted solutions violated the basic principles of analytical chemistry, even though other testing techniques, besides the counting of degranulated basophils, had been developed. Be that as it may, if a method is proposed for detecting trace amounts of cadmium, and it always yields positive results after the metal ions have been diluted out of the solution, it will justifiably be rejected as inadequate, subject to secondary effects that are responsible for the apparent results. The simple application of scientific rigor is the foundation on which analytical chemistry is based. Successive dilution is the means for measuring the sensitivity of the analytical methods and determining the threshold for the detection of the substance to be assayed. The use of strange detection systems is a characteristic of the discoveries of "pathological science." Piccardi's use of the sedimentation rate of bismuth oxychloride, a semi-ionic compound in which the metal is bound in an unusual polymeric structure layered with oxygen, is a typical example of this type of system, further reinforced by the turbulent conditions of the liquid mixture giving rise to the precipitate.

EPILOGUE

The "memory of water" can be considered as an extreme example, at the limit of fantasy, of the belief in the existence of a characteristic "structure" inherent in liquid water. *We have seen that if "structures" actually do exist, their duration is very brief*: it is impossible for them to exist for several hours, much less several days. Even in the absence of perturbation, they are destroyed by the perpetual dynamics of molecular motion. Even if the "coherent dissipative structures" recently proposed to account for the quantum delocalization of electrical conductivity by protons in water do exist—affecting perhaps thousands of molecules at a time, they are nevertheless destroyed on a time scale of 10^{-12} second. And even if they did "reproduce," the number of successive generations would undoubtedly very rapidly lead to "mutants" quite different from the form of the initial configuration.

Nevertheless, the question of the role of water in biochemical reactions still looms large, because it is a fundamental question. In the future, we hope that the discreet or the spectacular controversies that have always surrounded questions in physics of the chemistry of water and have stimulated research will contribute to the further advancement of science. As for the Water Goddesses, they will see to it that public interest is maintained in the subject.

155